Problems and Solutions in Electronics

Problems and Solutions in Electronics

Roger Loxton

CHAPMAN & HALL
University and Professional Division
London · Glasgow · Weinheim · New York · Tokyo · Melbourne · Madras

Published by Chapman & Hall, 2-6 Boundary Row, London SE1 8HN, UK

Chapman & Hall, 2-6 Boundary Row, London SE1 8HN, UK

Blackie Academic & Professional, Wester Cleddens Road, Bishopbriggs, Glasgow G64 2NZ, UK

Chapman & Hall GmbH, Pappelallee 3, 69469 Weinheim, Germany

Chapman & Hall USA, One Penn Plaza, 41st Floor, New York, NY10119, USA

Chapman & Hall Japan, ITP - Japan, Kyowa Building, 3F, 2-2-1 Hirakawacho, Chiyoda-ku, Tokyo 102, Japan

Chapman & Hall Australia, Thomas Nelson Australia, 102 Dodds Street, South Melbourne, Victoria 3205, Australia

Chapman & Hall India, R. Seshadri, 32 Second Main Road, CIT East, Madras 600 035, India

First edition 1994
Reprinted 1995

© 1994 Roger Loxton

Printed in Great Britain by The Alden Press, Oxford

ISBN 0 412 57820 4

Apart from any fair dealing for the purposes of research or private study, or criticism or review, as permitted under the UK Copyright Designs and Patents Act, 1988, this publication may not be reproduced, stored, or transmitted, in any form or by any means, without the prior permission in writing of the publishers, or in the case of reprographic reproduction only in accordance with the terms of the licences issued by the Copyright Licensing Agency in the UK, or in accordance with the terms of licences issued by the appropriate Reproduction Rights Organization outside the UK. Enquiries concerning reproduction outside the terms stated here should be sent to the publishers at the London address printed on this page.
   The publisher makes no representation, express or implied, with regard to the accuracy of the information contained in this book and cannot accept any legal responsibility or liability for any errors or omissions that may be made.

A Catalogue record for this book is available from the British Library

Library of Congress Cataloging-in-Publication Data available

∞ Printed on acid-free text paper, manufactured in accordance with ANSI/NISO Z39.48-1992 and ANSI/NISO Z39.48-1984 (Permanence of Paper).

CONTENTS

Preface		**vii**
1	**D.C. circuits and methods of circuit analysis**	**1**
	Questions	1
	Solutions	5
2	**Signals, waveforms and a.c. components**	**19**
	Questions	19
	Solutions	23
3	**Phasor analysis of a.c. circuits**	**31**
	Questions	31
	Solutions	37
4	**Amplifiers and feedback**	**55**
	Questions	55
	Solutions	60
5	**Combinational logic circuits**	**71**
	Questions	71
	Solutions	77
6	**Sequential logic circuits**	**95**
	Questions	95
	Solutions	101
7	**Analogue-digital conversion**	**117**
	Questions	117
	Solutions	120
8	**Memories, microprocessors and microcontrollers**	**129**
	Questions	129
	Solutions	132
9	**Diodes and transistors**	**137**
	Questions	137
	Solutions	139
10	**Analogue transistor circuits**	**145**
	Questions	145
	Solutions	150

Contents

11	**Basic transistor digital circuits**	**167**
	Questions	167
	Solutions	171
12	**Integrated-circuit technology**	**183**
13	**Power amplifiers**	**185**
	Questions	185
	Solutions	189
14	**Regulated d.c. power supplies**	**201**
	Questions	201
	Solutions	205
15	**Waveform generation**	**221**
	Questions	221
	Solutions	223
16	**The high-frequency behaviour of transistors**	**231**
	Questions	231
	Solutions	234
17	**Interconnections**	**243**
	Questions	243
	Solutions	246

PREFACE

This book of problems with worked solutions is designed to be used in connection with the study of the book **Electronics** by *Crecraft, Gorham and Sparkes*, published by Chapman & Hall (1993) [ISBN 0 412 41320 5 (0 442 30880 9 in USA)]. Throughout this work, that book is referred to for convenience as "the textbook".

The chapter numbers in this book correspond exactly with the chapter numbers in the textbook, which explains why this book has no content to its Chapter 12. Chapter 12 of the textbook is concerned with the technology of the manufacture of integrated circuits, and as such has no objectives relating to problem solving. This book therefore contains no problems relating to the material of that chapter. Some chapters of the textbook contain more problem-solving related objectives than others, and because of that some chapters of this book contain more problems than others.

The problems contained in this book have been chosen to serve one or more of several purposes. The first purpose is to provide the opportunity for the reader to practise skills which have been taught in the textbook, particularly where I have judged a skill as being difficult to acquire. The second purpose is to require the student to integrate techniques from more than one chapter of the textbook in solving a problem, something which the textbook self-assessment questions do not generally attempt. A third purpose is to use a problem and the wording of its solution to clarify a point which I consider potentially confusing in the textbook. An occasional fourth purpose has been to highlight some aspect of the material which I consider important, but which has not been specifically highlighted in the textbook.

Do not expect to find a direct parallel to each question posed here in the self-assessment questions in the textbook. Most of these questions will prove more challenging than those in the textbook, not only because some require you to integrate different aspects of the textbook material, but because they have been designed to require you to think about *how* to apply the theory to the specific problem in hand.

The solutions to the questions are fully worked, and, where necessary, contain explanations of why the approach taken to obtain the solution is the most appropriate. They even, occasionally, contain some teaching where it has seemed appropriate or necessary.

Having said all that, it is also true to say that all the theory and method required to solve these problems is contained in the textbook. You are not required to search other texts in order to be equipped to answer them.

When solving a problem in Electronics, or designing a circuit or system, it is normal good practice nowadays to test a solution by simulating the circuit or system using an appropriate software package. Regrettably, owing to lack of time and access to appropriate software at the right time, the solutions given here have not been so checked. If you decide to check them and find any problems, please do not hesitate to point them out to me via the publishers so that future readers can have the benefit of more fully tried and tested solutions.

From time to time this book refers to figures in the textbook, and this occasionally creates a potential for confusion between figures in this book and figures in the textbook. *Unless specifically stated otherwise, all figure references in problems and their solutions are to figures in* **this** *book, not the textbook.*

SYMBOLS USED IN THIS BOOK

Symbols are listed under the heading of the chapter in which they are first used. Greek symbols are listed after English ones. The values of constant quantities are given.

Chapter 1

I	Current (d.c., a.c. amplitude or a.c. r.m.s. value)	ampere
I_N	Norton equivalent circuit current source	ampere
R	D.C. resistance of a resistor	ohm
R_N	Norton equivalent circuit resistance	ohm
R_T	Thévenin equivalent circuit resistance	ohm
V	Voltage (d.c., a.c. amplitude or a.c. r.m.s. value)	volt
V_T	Thévenin equivalent circuit voltage source	volt

Chapter 2

C	Capacitance	farad
f	Frequency	hertz
i	Current, instantaneous value	ampere
L	Inductance	henry
t	Time	second
V_s	Supply voltage or source voltage	volt
v	Voltage, instantaneous value	volt
X	Reactance	ohm
ε (epsilon)	Electric permittivity ($\varepsilon = \varepsilon_0 \varepsilon_r$)	F m^{-1}
ε_0	Permittivity of free space (= 8.854)	pF m^{-1}
ε_r	Relative permittivity	dimensionless
ω (omega)	Angular frequency ($\omega = 2\pi f$)	radians s^{-1}

Chapter 3

A	Voltage transfer function (phasor operator)	dimensionless
I	Current phasor	ampere
T	Time constant	second
V	Voltage phasor	volt
Y	Admittance (phasor operator)	siemens
Z	Impedance (phasor operator)	ohm
ϕ (phi)	Phase angle	radian

Chapter 4

A_v	Low-frequency voltage gain of an amplifier	dimensionless
A$_v$	Voltage gain of an amplifier (phasor operator)	dimensionless
G	Low-frequency closed-loop gain of a feedback amplifier	dimensionless
G	Closed-loop gain of a feedback amplifier (phasor operator)	dimensionless
i_{NA}	Equivalent input noise current of an amplifier	ampere
I_B	Input bias current of an amplifier	ampere
I_{IO}	Input offset current of a differential amplifier	ampere
I_{NA}	Equivalent input noise current of an amplifier (r.m.s.)	ampere

Symbols

r	Slope resistance or small-signal resistance	ohm
v_{NA}	equivalent input noise voltage of an amplifier	volt
V_{IO}	Input offset voltage of a diffferentiual amplifier	volt
V_{NA}	Equivalent input noise voltage of an amplifier (r.m.s.)	volt
$\mathbf{z}$	Small-signal impedance (phasor operator)	ohm
β (beta)	Feedback ratio	dimensionless

Chapter 9

I_D	pn junction diode current	ampere
I_S	Saturation current of a pn junction	ampere
I_{SE}	Saturation current of the emitter-base junction of a bipolar transistor	ampere
k	Boltzmann's constant ($= 1.3806 \times 10^{-23}$)	J K^{-1}
K	Constant $= q/kT \approx 40$ at $T = 293$ K	V^{-1}
n_i	Intrinsic carrier density in a semiconductor	dimensionless
n_{p0}	Equilibrium electron density in p-type semiconductor	dimensionless
n_0	Equilibrium density of electrons in a semiconductor	dimensionless
N_a, N_d	Acceptor, donor density in a semiconductor	dimensionless
p_{n0}	Equilibrium hole density in n-type semiconductor	dimensionless
p_0	Equilibrium density of holes in a semiconductor	dimensionless
q	Electron charge ($= 1.6 \times 10^{-19}$)	coulomb
T	Temperature	kelvin
VA	Early voltage in a bipolar transistor	volt
V_D	D.C. voltage applied to a diode	volt
α (alpha), α_N	D.C. current ratio of a bipolar transistor (I_C/I_E)	dimensionless
β (beta), β_N	D.C. current ratio of a bipolar transistor (I_C/I_B)	dimensionless
μ_n (mu)	Electron mobility in silicon	m^2 V^{-1} s^{-1}
μ_p	Hole mobility in silicon	m^2 V^{-1} s^{-1}
ρ (rho)	Resistivity	Ω m
σ (sigma)	Conductivity	S m^{-1}
σ_n, σ_p	Conductivity of n-region, p-region of silicon	S m^{-1}

Chapter 10

g	Small-signal conductance	siemens
g_i, g_o	Input, output conductance	siemens
g_m	Mutual conductance or transconductance	A V^{-1}
r_i, r_o	input, output slope resistance of transistor	ohm
V_{CC}	D.C. supply voltage	volt
τ (tau)	Time constant	second

Chapter 11

I_{CBO}	Common-base collector cut-off current	ampere
I_{CEO}	Common-emitter collector cut-off current	ampere
Q	Charge	coulomb
Q_B	Minority carrier charge in the base region of a bipolar transistor	coulomb
Q_{BS}	Saturation charge in the base region	coulomb

Symbols

Q_{off}	Charge required to turn off a transistor switch	coulomb
Q_{on}	Charge required to turn on a transistor switch	coulomb
t_{on}, t_{off}	Turn-on, turn-off time of a transistor switch	second
t_d, t_f, t_r, t_s	Delay time, fall time, risetime, saturation time of a transistor switch	second
V_T	Threshold voltage of a MOSFET	volt
α_I	Inverse d.c. current ratio (I_E/I_C)	dimensionless
β	Gain factor of a MOSFET	dimensionless
λ (lambda)	Channel length modulation factor (MOSFETs)	V^{-1}
τ_t	Base transit time of a bipolar transistor	second
τ_s	Saturation time constant of a bipolar transistor	second

Chapter 13

I_Q	Quiescent output transistor current	ampere

Chapter 14

Δ (u.c. delta)	A finite increment	

Chapter 16

C_c	Collector capacitance	farad
C_i	Transistor input capacitance	farad
C_{de}	Emitter diffusion capacitance	farad
C_{et}	Emitter transition-region capacitance	farad
f_b	Common-emitter current-gain cut-off frequency	hertz
f_T	Transition frequency	hertz
f_1	Frequency at which $|\mathbf{h}_{fe}| = 1$	hertz
h_{fe}	Low-frequency, small-signal common-emitter curent gain	dimensionless
$\mathbf{h}_{fe}$	High-frequency, small-signal common-emitter curent gain	dimensionless
r_b	Extrinsic base resistance of a bipolar transistor	ohm
ϕ (phi)	Contact potential of a pn junction	volt

Chapter 17

c	Speed of light and other electromagnetic waves	m s^{-1}
v	Propagation velocity	m s^{-1}
$\mathbf{Z}_0$	Characteristic impedance of a line	ohm
α	Attenuation per unit length of a transmission line	neper m^{-1}
Γ (u.c. gamma)	Voltage reflection coefficient of a transmission line	dimensionless

1 D.C. CIRCUITS AND METHODS OF CIRCUIT ANALYSIS

QUESTIONS

1.1 For the circuit of Figure 1.31 (page 20) of the textbook, write down the Kirchhoff voltage law equations for loops FABCGEF and EDBCGE.

1.2 (a) Use Thévenin's theorem to obtain the Thévenin equivalent of the circuit contained in the dotted box in Figure 1.1. Hence calculate the voltage across a 1 kΩ load resistor.

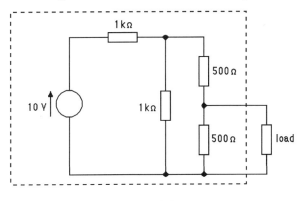

Figure 1.1

(b) Use Norton's theorem to obtain the Norton equivalent of the circuit contained in the dotted box in Figure 1.1. Hence calculate the current through a 1 kΩ load resistor.

(c) Confirm that the answers obtained in parts (a) and (b) are compatible.

1.3 A moving coil multimeter has a coil resistance of 500 Ω and a full-scale deflection current of 0.5 mA.

(a) What value of resistance must be connected in series with the coil to provide a full-scale deflection voltage of 100 V?

(b) What value of resistance must be connected in parallel with the coil to provide a full-scale deflection current of 1 A?

(c) What is the "ohms-per-volt" figure for the meter?

1.4 (a) Using any appropriate method, calculate the voltage across resistor R_1 in the circuit of Figure 1.2.

(b) The voltage across R_1 is measured using the 10 volt f.s.d. range of a 5 000 ohms-per-volt moving-coil multimeter. Assuming negligible errors in the meter, what will be the meter reading?

D.C. circuits and methods of circuit analysis

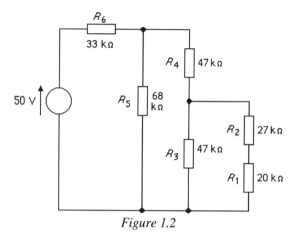

Figure 1.2

(c) What percentage error has been introduced into the reading by the measurement method?

(d) What will be the percentage error if a 20 000 Ω/V moving-coil meter is used instead? (Assume that the full-scale deflection voltage is still 10 V.)

1.5 (a) Calculate the current flowing in the 1 Ω resistor of Figure 1.3.

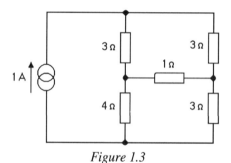

Figure 1.3

(b) This current is measured, using the same multimeter as in Question 1.4(b), switched to a current range of 100 mA. The meter coil resistance is 1 kΩ and a shunt resistor is used to establish each current range. Assuming that negligible error exists in the meter, what will be the meter reading?

(c) What percentage error has been introduced into the reading by the measurement method?

1.6 (a) Figure 1.4(a) is the circuit of a Wheatstone bridge supplied from an ideal voltage source. Use Thévenin's theorem to calculate the component values of the Thévenin equivalent circuit for the bridge. Hence calculate the output voltage from the bridge when loaded with a resistance of 1 kΩ.

(b) Figure 1.4(b) is the circuit of the same bridge as in part (a), but supplied from an ideal current source. Again calculate the Thévenin equivalent circuit values.

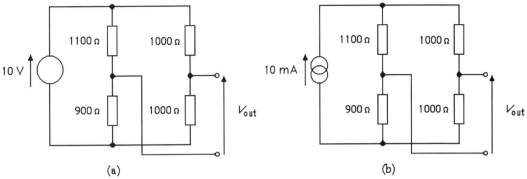

(a) (b)

Figure 1.4

(c) Without performing any calculations, deduce the approximate Thévenin equivalent resistance of the bridge arrangement of Figure 1.5.

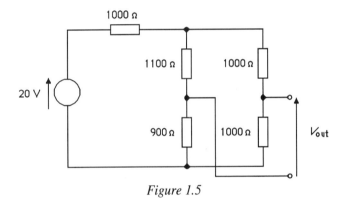

Figure 1.5

1.7 Using Norton's theorem together with the superposition principle, calculate the Norton equivalent circuit component values for the circuit of Figure 1.6. Hence deduce the current that will flow in a load resistance of 330 Ω.

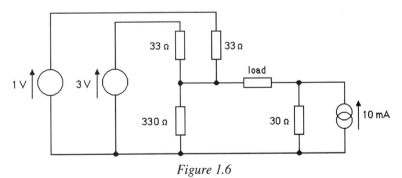

Figure 1.6

D.C. circuits and methods of circuit analysis

1.8 (a) Use nodal analysis to obtain the values of all the node voltages in the circuit of Figure 1.7.

(b) The right hand 4 kΩ resistor of the circuit of Figure 1.7 is replaced with a 10 mA ideal current source acting upwards. Using nodal analysis, re-calculate the node voltages.

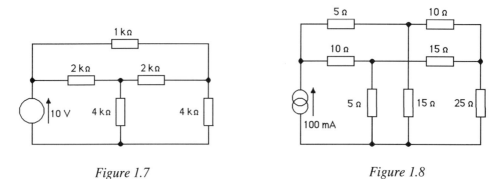

Figure 1.7   Figure 1.8

1.9 Use nodal analysis to obtain the simultaneous equations specifying the node voltages of Figure 1.8.

1.10 Figure 1.9 represents a "cube" structure of 12 resistors, each resistor having the same value of 1 kΩ. A constant current source of 60 mA is connected across opposite corners of the "cube". Calculate the voltage across the current source.

(**Hint:** rather than having to perform complex calculations or solve a large number of simultaneous equations to evaluate each branch current, intuitive thinking can lead to a solution with minimal calculation.)

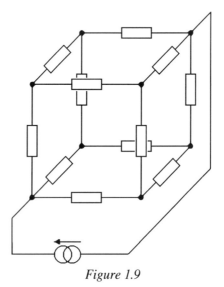

Figure 1.9

SOLUTIONS

Question 1.1

For loop FABCGEF, the Kirchhoff voltage law equation is:

$$V_1 - I_1 R_1 - I_4 R_4 - I_7 R_7 - I_8 R_8 - V_2 = 0$$

while for loop EDBCGE the equation is:

$$I_5 R_5 + I_6 R_6 - I_7 R_7 - I_8 R_8 - V_2 = 0$$

Question 1.2

(a)

To obtain the Thévenin equivalent resistance R_T of the circuit contained in the dotted box in Figure 1.1, we first replace the voltage source with its internal resistance, which in this case is zero. Redrawing the figure gives the arrangement of Figure 1.10, from which you can see that the equivalent resistance of the two 1 kΩ resistors in parallel is 500 Ω, so that the value of R_T is (500 + 500) Ω in parallel with 500 Ω, giving:

$$R_T = \frac{500 \times 1000}{1500}\,\Omega = \frac{1000}{3}\,\Omega$$

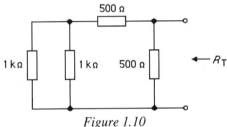

Figure 1.10

To obtain the Thévenin equivalent voltage, the load is removed, so that the open-circuit voltage V_T can be calculated. In this case, V_T will be exactly one-half of the voltage V_1 in Figure 1.11.

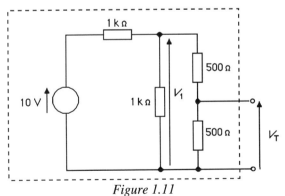

Figure 1.11

D.C. circuits and methods of circuit analysis

V_1 can be calculated as the voltage across 1 kΩ in parallel with (500 + 500) Ω, i.e. across 500 Ω.

Hence: $$V_1 = 10 \text{ V} \times \frac{500}{1500} = \frac{10}{3} \text{ V}$$

and so, $$V_T = \frac{1}{2} \times \frac{10}{3} \text{ V} = \frac{5}{3} \text{ V}$$

The circuit of Figure 1.1 can therefore be represented by the Thévenin equivalent shown in Figure 1.12.

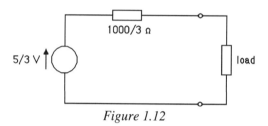

Figure 1.12

When the load is 1 kΩ, the voltage across the load is given by:

$$V_L = \frac{5}{3} \text{ V} \times \frac{1000}{1000 + (1000/3)} = \frac{5}{3} \text{ V} \times \frac{3}{4} = 1.25 \text{ V}$$

(b)

For the Norton equivalent circuit, R_N will be equal to R_T of part (a), so $R_N = 1000/3$ Ω

I_N can be found from the circuit of Figure 1.13(a), and a convenient way of finding the value of I_N is to use a Thévenin equivalent for part of the circuit as shown in Figure 1.13(b). From this figure it can easily be seen that, since no current will flow in the rightmost 500 Ω resistor because of the short-circuit across it, $I_N = 5$ mA.

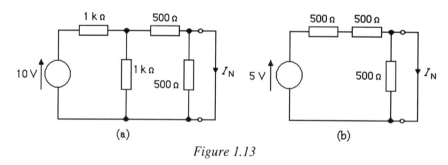

Figure 1.13

The Norton equivalent circuit is shown in Figure 1.14, and the value of the current in the 1 kΩ load can be calculated using the current divider rule as:

$$I = 5 \text{ mA} \times \frac{1000/3}{1000 + 1000/3} = 1.25 \text{ mA}$$

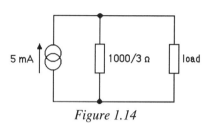

Figure 1.14

(c)

These two results are entirely compatible, since 1.25 mA flowing through 1 kΩ produces a voltage drop of 1.25 V.

Question 1.3

(a)

For a full-scale deflection voltage of 100 V, the total resistance of meter coil and series resistor must be 100 V ÷ 0.5 mA = 200 kΩ. The value of the series resistor is therefore 200 kΩ – 500 Ω = 195.5 kΩ.

(b)

For a full-scale deflection current of 1 A, the current through the shunt resistor must be 999.5 mA when the current through the meter is 0.5 mA. The resistance of the shunt must therefore equal the resistance of the coil divided by 999.5/0.5. Hence the value of the shunt resistor is 500 × 0.5 ÷ 995.5 Ω = 0.25 Ω.

(c)

The meter has an "ohms-per-volt" figure equal to the reciprocal of the full-scale deflection current, i.e. 1/(0.5 mA) = 2 000 Ω/V.

Question 1.4

There are a number of methods which can be used to obtain the solution to this question. For example, referring to Figure 1.2, the voltage across R_1 can be calculated as a proportion of the voltage across R_3, which can be calculated as a proportion of the voltage across R_5, which in turn can be found as a proportion of the source voltage. An alternative is to use nodal analysis to obtain equations defining the node voltages and then solve these equations to obtain the required answer. Another alternative is to use either Norton's theorem or Thévenin's theorem to obtain the equivalent circuit of everything but resistor R_1, and hence calculate the voltage across R_1.

In this question, the clue as to which is the best method to use is given by the other parts of the question, where parts (b) and (d) clearly indicate the need to repeat the calculation with an altered value of R_1. With either the first method suggested above, or the use of nodal analysis (unless, of

D.C. circuits and methods of circuit analysis

course, you are using a software package), changing the value of R_1 means repeating each part of the calculation for the new value. However, with either Norton's theorem or Thévenin's theorem, since all but the final step of the calculation is made with R_1 removed, only the final step will need repeating for each new value of R_1. Whether you use Norton's theorem or Thévenin's theorem is a matter of personal choice, and I have chosen to use Thévenin's theorem.

(a)

To simplify the creation of the Thévenin equivalent circuit, I have chosen to apply Thévenin's theorem in steps to parts of the circuit. To start with, I am going to find the equivalent circuit for the source, R_6, R_5 and R_4 with R_3, R_2 and R_1 removed.

The equivalent circuit component values will be given by:

$$R_{T1} = 47 \text{ k}\Omega + \frac{33 \times 68}{33 + 68} \text{ k}\Omega = 69.2 \text{ k}\Omega$$

$$V_{T1} = 50 \text{ V} \times \frac{68}{33 + 68} = 33.7 \text{ V}$$

(since there is no voltage drop across R_4).

Using these values gives the equivalent circuit of Figure 1.15.

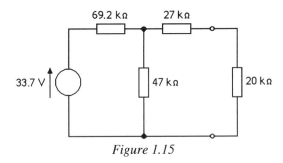

Figure 1.15

Removing the 20 kΩ resistor (R_1) allows the Thévenin equivalent for the whole circuit (less R_1) to be obtained.

$$R_{T2} = 27 \text{ k}\Omega + \frac{47 \times 69.2}{47 + 69.2} \text{ k}\Omega = 55 \text{ k}\Omega$$

$$V_{T2} = 33.7 \text{ V} \times \frac{47}{47 + 69.2} = 13.6 \text{ V}$$

The voltage across R_1 is then given by:

$$V_{R_1} = 13.6 \text{ V} \times \frac{20}{20 + 55} = 3.63 \text{ V}$$

(b)

When a 5 000 Ω/V meter on its 10 V range is connected across R_1, the value of the combination of resistance and meter is 20 kΩ in parallel with 50 kΩ = 14.3 kΩ. Using the same values of R_{T_2} and V_{T_2} as in part (a), the value of the voltage across R_1 becomes:

$$V_{R_1} = 13.6 \text{ V} \times \frac{14.3}{14.3 + 55} = 2.81 \text{ V}$$

This, therefore will be the meter reading.

(c)

The percentage error introduced by the meter will be:

$$E = \frac{3.63 - 2.81}{3.63} \times 100\% = 23\%$$

(d)

If a 20 000 Ω/V meter is used to measure the voltage across R_1, the new value of the combined resistance will be 200 kΩ in parallel with 20 kΩ = 18.2 kΩ.

The new value of the voltage across R_1 will then be:

$$V_{R_1} = 13.6 \text{ V} \times \frac{18.2}{18.2 + 55} = 3.38 \text{ V}$$

and the new value of the percentage error will be:

$$E = \frac{3.63 - 3.38}{3.63} \times 100\% = 6.9\%.$$

Question 1.5

Again, there are a number of different methods for calculating the value of the required current. Again, as in Question 1.4, the requirement to re-calculate the value of the current with a changed value of resistance for part (b) suggests that either a Thévenin or a Norton equivalent circuit will provide the most convenient method. I have again chosen to use the Thévenin equivalent circuit because, in this question, the open-circuit voltage is easier to calculate than the short-circuit current.

(a)

The open-circuit voltage of the circuit, with the 1 Ω resistor considered as the load on the circuit, can be found by calculating the difference $V_B - V_A$ in Figure 1.16.

Using the current divider rule,

$$I_1 = 7/(6 + 7) \text{ A} = 7/13 \text{ A}$$

$$I_2 = 6/(6 + 7) \text{ A} = 6/13 \text{ A}$$

D.C. circuits and methods of circuit analysis

So, $V_T = V_B - V_A = 4\,\Omega \times 6/13\,\text{A} - 3\,\Omega \times 7/13\,\text{A} = (24/13 - 21/13)\,\text{V} = 3/13\,\text{V}.$

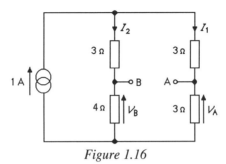

Figure 1.16

To calculate the value of R_T, replace the current source with its infinite internal resistance, giving:

$$R_T = 6\,\Omega \text{ in parallel with } 7\,\Omega = 6 \times 7 /(6 + 7)\,\Omega = 42/13\,\Omega.$$

The value of the current in the 1 Ω resistor connected between A and B can then be calculated as:

$$I = \frac{V_T}{R_T + 1\,\Omega} = \frac{3/13\,\text{V}}{(42/13 + 1)\,\Omega} = \frac{3}{55}\,\text{A} = 54.5\,\text{mA}$$

(b)

If the moving coil meter has a coil resistance of 1000 Ω and full-scale deflection current of 0.2 mA (the reciprocal of the "ohms-per-volt" figure of 5000), the voltage drop across it at full-scale current must be 0.2 V. Hence, when shunted to provide a full-scale deflection current of 100 mA, the total resistance of meter plus shunt must be 0.2 V ÷ 100 mA = 2 Ω.

To measure the current in the 1 Ω resistor, the meter is connected in series with the resistor, so increasing the resistance of that branch of the circuit from 1 Ω to 3 Ω.

The new value of the current can then be calculated as:

$$I = \frac{V_T}{R_T + 3} = \frac{3/13\,\text{V}}{(42/13 + 3)\,\Omega} = \frac{3}{81}\,\text{A} = 37.0\,\text{mA}$$

and this will be the meter reading.

(c)

The percentage error introduced by the measurement method can be calculated as:

$$E = \frac{54.5 - 37.0}{54.5} \times 100\% = 32\%$$

Solutions

Question 1.6

(a)

As in the previous question, the value of V_T can be found as the difference between the voltages at the two output terminals. So,

$$V_T = 10\text{ V} \times \frac{1000}{1000+1000} - 10\text{ V} \times \frac{900}{900+1100} = 5\text{ V} - 4.5\text{ V} = 0.5\text{ V}$$

Replacing the ideal voltage source with its zero internal resistance gives the value of R_T as:

$$R_T = \frac{1000 \times 1000}{1000+1000}\,\Omega + \frac{900 \times 1100}{900+1100}\,\Omega = 500\,\Omega + 495\,\Omega = 995\,\Omega$$

When the bridge is loaded with a resistance of 1 kΩ, the output voltage will have the value:

$$V_O = V_T \times \frac{1\text{ k}\Omega}{R_T + 1\text{ k}\Omega} = 0.5\text{ V} \times \frac{1}{1.995} = 0.25\text{ V}$$

(b)

Because both halves of the bridge have equal total resistance of 2 kΩ, the current from the source will divide equally between them. The Thévenin equivalent voltage can therefore be calculated as:

$$V_T = 5\text{ mA} \times 1000\,\Omega - 5\text{ mA} \times 900\,\Omega = 0.5\text{ V}$$

Replacing the current source with its infinite internal resistance gives the Thévenin equivalent resistance as (1000 + 1100) Ω in parallel with (1000 + 900) Ω; i.e.

$$R_T = \frac{2100 \times 1900}{2100 + 1900}\,\Omega = 997.5\,\Omega$$

(c)

Since the Thévenin equivalent resistance is almost exactly 1000 Ω with both a zero resistance voltage source and an infinite resistance current source, the Thévenin resistance must be practically independent of the resistance of the source of bridge energisation. Hence the Thévenin equivalent resistance of the circuit of Figure 1.5 will also be almost exactly 1000 Ω.

Question 1.7

Re-drawing the circuit as in Figure 1.17, the value of R_N can be seen to be 30 Ω plus the parallel combination of 33 Ω, 33 Ω, and 330 Ω.

The resistance of the parallel combination can be calculated by adding the conductances and then finding the reciprocal of the sum, i.e.

$$R = \frac{1}{(1/33 + 1/33 + 1/330)}\,\Omega = \frac{1}{(21/330)}\,\Omega = \frac{330}{21}\,\Omega = 15.7\,\Omega$$

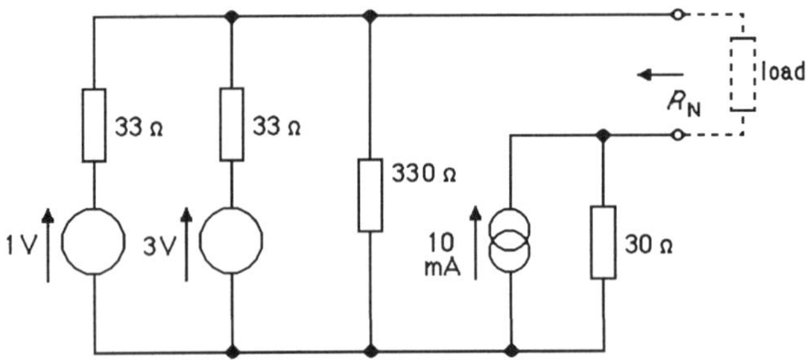

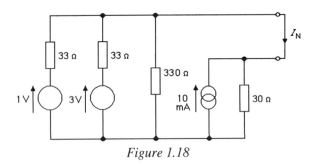

Figure 1.17

Hence, $R_N = (30 + 15.7)\,\Omega = 45.7\,\Omega$

I_N is the current which will flow in a short circuit across the output terminals of the network as shown in Figure 1.18.

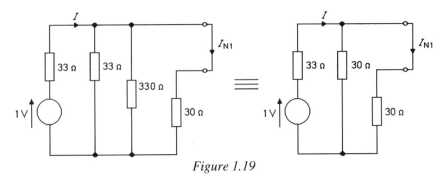

Figure 1.18

To find the value of this current we use the superposition principle, calculating the contribution to I_N of each source in turn, all other sources being replaced by their internal resistances.

First, the 1 V source alone (Figure 1.19).

Figure 1.19

Solutions

$$I_{N1} = \tfrac{1}{2}I = \tfrac{1}{2}\left(\frac{1\text{ V}}{33\,\Omega + 30/2\,\Omega}\right) = \frac{1}{96}\text{ A} = 10.4\text{ mA}$$

The contribution of the 3 V source will be just three times that of the 1 V source (since they are both in series with a resistance of 33 Ω), so

$$I_{N2} = 31.2\text{ mA}$$

The contribution of the 10 mA source (Figure 1.20) can be found using the current divider rule.

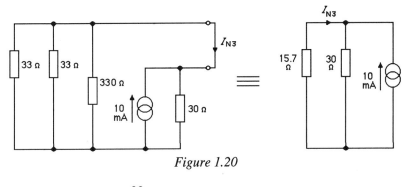

Figure 1.20

$$I_{N3} = -10\text{ mA} \times \frac{30}{(30+15.7)} = -6.56\text{ mA}$$

Adding these three contributions together,

$$I_N = I_{N1} + I_{N2} + I_{N3} = (10.4 + 31.2 - 6.56)\text{ mA} = 35.0\text{ mA}$$

The current through a 330 Ω load will therefore be (Figure 1.21):

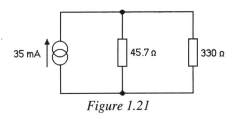

Figure 1.21

$$I = 35.0\text{ mA} \times \frac{45.7}{330 + 45.7} = 4.26\text{ mA}$$

Question 1.8

(a)

Figure 1.22 shows the circuit with node voltages and branch currents labelled.

D.C. circuits and methods of circuit analysis

At node 1, $I_2 = I_3 + I_4$ (1)

At node 2, $I_3 + I_1 = I_5$ (2)

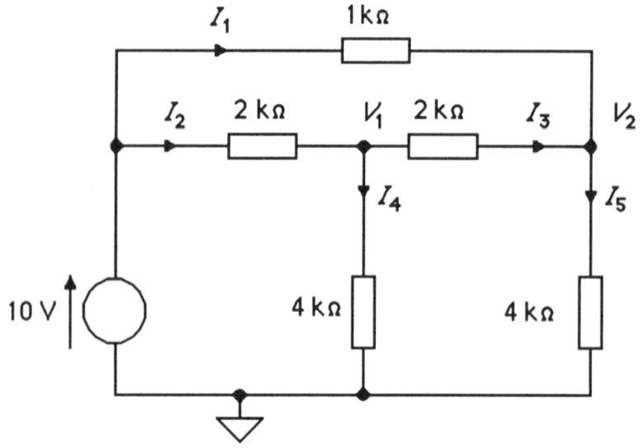

Figure 1.22

To make the equations clearer, assume all voltages are in volts, all currents in milliamps and all resistances in kilohms. The expressions for the currents in terms of node voltages can then be written:

$$I_1 = \frac{10 - V_2}{1}, \quad I_2 = \frac{10 - V_1}{2}, \quad I_3 = \frac{V_1 - V_2}{2}, \quad I_4 = \frac{V_1}{4}, \quad I_5 = \frac{V_2}{4}$$

Substituting these current expressions in equation (1),

$$\frac{10 - V_1}{2} = \frac{V_1 - V_2}{2} + \frac{V_1}{4}$$

$$5 = V_1 \left[\frac{1}{2} + \frac{1}{2} + \frac{1}{4} \right] - \frac{V_2}{2}$$

$$20 = 5V_1 - 2V_2 \quad (3)$$

and in equation (2),

$$\frac{V_1 - V_2}{2} + \frac{10 - V_2}{1} = \frac{V_2}{4}$$

$$10 = V_2 \left[\frac{1}{4} + 1 + \frac{1}{2} \right] - \frac{V_1}{2}$$

$$40 = 7V_2 - 2V_1 \quad (4)$$

Multiply equation (3) by 2 and equation (4) by 5 and add.

$$240 = 31V_2$$
$$V_2 = 7.74 \text{ V}$$

Substitute for V_2 in equation (3),

$$20 = 5V_1 - 15.48$$
$$V_1 = \frac{35.48}{5}$$
$$= 7.1 \text{ V}$$

(b)

Replacing the right-hand 4 kΩ resistor with a 10 mA current source gives Figure 1.23

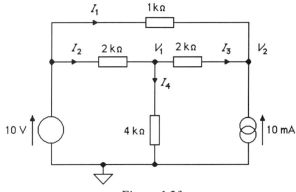

Figure 1.23

At node 1, $I_2 = I_3 + I_4$ as before.

At node 2, $I_3 + I_1 + 10 = 0$.

The expressions for currents I_1, I_2, I_3 and I_4 will be unchanged from part (a).

The node 1 equation will therefore be the same as in part (a), i.e.

$$20 = 5V_1 - 2V_2 \qquad (3)$$

The node 2 equation is:

$$\frac{V_1 - V_2}{2} + \frac{10 - V_2}{1} + 10 = 0$$

$$20 = V_2 \left[\frac{1}{2} + 1\right] - \frac{V_1}{2}$$

$$40 = 3V_2 - V_1 \qquad (4)$$

D.C. circuits and methods of circuit analysis

Solving these two equations in a similar manner to part (a) yields
$$V_1 = 10.8 \text{ V}, V_2 = 16.9 \text{ V}$$

Question 1.9

In the circuit of Figure 1.8 there will be 4 unknown node voltages as indicated in Figure 1.24

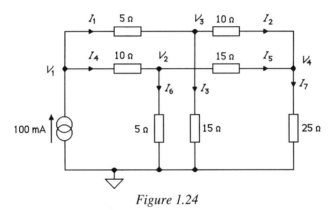

Figure 1.24

The node equations (all currents in mA) are:

$$100 = I_1 + I_4 \quad (1)$$
$$I_4 = I_5 + I_6 \quad (2)$$
$$I_1 = I_2 + I_3 \quad (3)$$
$$I_2 + I_5 = I_7 \quad (4)$$

The current expressions (with all voltages in mV) are:

$$I_1 = \frac{V_1 - V_3}{5}, \quad I_2 = \frac{V_3 - V_4}{10}, \quad I_3 = \frac{V_3}{15}, \quad I_4 = \frac{V_1 - V_2}{10},$$

$$I_5 = \frac{V_2 - V_4}{15}, \quad I_6 = \frac{V_2}{5}, \quad I_7 = \frac{V_4}{25}.$$

Equation (1) yields

$$100 = \frac{V_1 - V_3}{5} + \frac{V_1 - V_2}{10}$$

$$100 = V_1 \left[\frac{1}{5} + \frac{1}{10}\right] - \frac{V_2}{10} - \frac{V_3}{5}$$

$$1000 = 3V_1 - V_2 - 2V_3 \quad \text{(A)}$$

Equation (2) yields

$$\frac{V_1-V_2}{10} = \frac{V_2-V_4}{15}+\frac{V_2}{5}$$

$$0 = V_2\left[\frac{1}{10}+\frac{1}{15}+\frac{1}{5}\right]-\frac{V_1}{10}-\frac{V_4}{15}$$

$$0 = 11V_2 - 3V_1 - 2V_4 \tag{B}$$

Equation (3) yields

$$\frac{V_1-V_3}{5} = \frac{V_3-V_4}{10}+\frac{V_3}{15}$$

$$0 = V_3\left[\frac{1}{10}+\frac{1}{15}+\frac{1}{5}\right]-\frac{V_1}{5}-\frac{V_4}{10}$$

$$0 = 11V_3 - 6V_1 - 3V_4 \tag{C}$$

Equation (4) yields

$$\frac{V_3-V_4}{10}+\frac{V_2-V_4}{15} = \frac{V_4}{25}$$

$$0 = V_4\left[\frac{1}{10}+\frac{1}{15}+\frac{1}{25}\right]-\frac{V_2}{15}-\frac{V_3}{10}$$

$$0 = 31V_4 - 10V_2 - 15V_3 \tag{D}$$

Question 1.10

Figure 1.25 is Figure 1.9 with the branch currents identified.

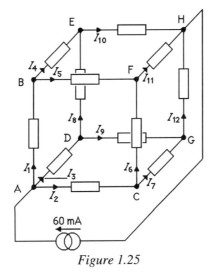

Figure 1.25

D.C. circuits and methods of circuit analysis

Because of the symmetry of the circuit, the values of the branch currents can be easily deduced. For example, at node A, the source current will divide equally between the three resistors connected to that node, so that the currents I_1, I_2 and I_3 will each be 20 mA.

Similarly, at each of nodes B, C and D, the current entering the node will divide equally between the other two branches connected to the node, so that I_4, I_5, I_6, I_7, I_8 and I_9 will each be 10 mA.

At nodes E, F and G these currents recombine so that I_{10}, I_{11} and I_{12} are each 20 mA. Finally, at node H, these three currents combine to form the current of 60 mA flowing through the source.

The value of the voltage across the source can be evaluated by considering any one of the parallel paths between nodes A and H, for example the path ABFH. The voltage drop along this path will be:

$$V = 1\,\text{k}\Omega \times 20\,\text{mA} + 1\,\text{k}\Omega \times 10\,\text{mA} + 1\,\text{k}\Omega \times 20\,\text{mA}$$
$$= 50\,\text{V}.$$

2 SIGNALS, WAVEFORMS AND A.C. COMPONENTS

QUESTIONS

2.1 Figure 2.1 shows a square-wave voltage which is symmetrical about the time axis.

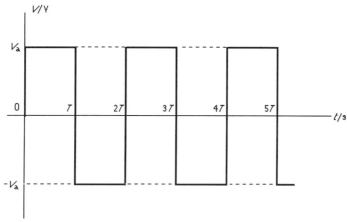

Figure 2.1

(a) What are the mean and r.m.s. values of the waveform?

(b) The voltage is measured on a rectifying moving-coil meter, calibrated to indicate the r.m.s. value of a sinusoidal input voltage. What error will there be in the indication of the r.m.s. value of the square-wave because of its non-sinusoidal nature, assuming that the meter itself introduces no significant error into the measurement?

2.2 A single cycle of a repetitive voltage waveform can be expressed mathematically as:

$V = (10 - 4t)$ V   for t between 0 and 1 seconds

$V = -10 + 4(t - 1)$ V   for t between 1 and 2 seconds.

(a) Sketch the waveform.

(b) Calculate its mean and r.m.s. values.

2.3 The following samples of a non-sinusoidal periodic voltage waveform were taken at 1 ms intervals and cover the whole of one cycle of the waveform.

(a) Find the mean and r.m.s. values of the waveform.

(b) What is the fundamental frequency of the waveform?

Samples (in volts):   0, 1.2, 1.8, 2.0, 2.1, 2.9, 5.4, 6.8, 6.3, 4.3, 1.0, –1.5, –2.8, –2.9, –2.7, –3.4, –4.5, –6.1, –7.4, –7.2, –4.6, –2.3, –0.9.

Signals, waveforms and a.c. components

2.4 The Fourier series for the triangle voltage waveform of Figure 2.2 can be expressed in the form

$$v = k_0 + k_1 \sin \omega_0 t + k_2 \sin 2\omega_0 t + k_3 \sin 3\omega_0 t + \ldots$$

where $f_0 = \omega_0/2\pi$ is the fundamental frequency, and the value of k_1, k_2, k_3, etc. can be found by substituting values of $n = 1, 2, 3$, etc. into the expression:

$$k_n = V_a \left(\frac{\sin(n\pi/2)}{n\pi/2} \right)^2$$

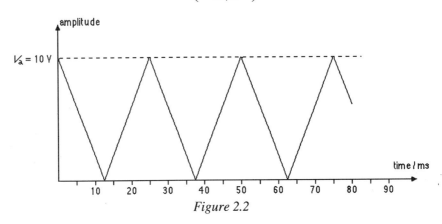

Figure 2.2

(a) Using this expression, calculate the first 6 terms of the Fourier series and draw these components as a line frequency spectrum.

(b) Which is the first non-zero harmonic with an amplitude less than 10 mV? (You do *not* have to calculate the amplitude of each harmonic until you find one less than 10 mV.)

2.5 A music signal is filtered to produce a signal having the power density spectrum shown in Figure 2.3. What is the r.m.s. voltage of the signal?

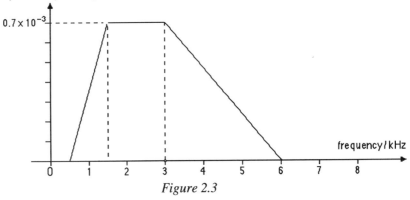

Figure 2.3

Questions

2.6 A capacitor is constructed from two conducting surfaces, each having dimensions 20 mm × 250 mm, separated by a dielectric of relative permittivity $\varepsilon_r = 350$ and thickness 15 μm.

  (a) What is the capacitance of the capacitor?

  (b) What is the reactance of the capacitor at 2 kHz?

  (c) What current will flow in the capacitor when it is connected across an a.c. supply of 240 V at 50 Hz?

2.7 **(a)** What is the time constant of the circuit of Figure 2.4?

  (b) Assuming that the capacitor is initially uncharged, what will be the value of the output voltage 4.5 seconds after the switch position is changed from 1 to 2?

  (c) What will be the values of the output voltage and the capacitor voltage after 15 seconds?

  (d) After 15 seconds, the switch is returned to position 1. Sketch the variation of output voltage with time after this change, annotating both axes appropriately.

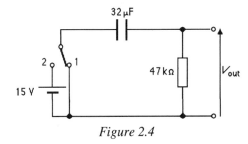

Figure 2.4

2.8 **(a)** Use Thévenin's theorem to find the time constant of the charging of the capacitor in Figure 2.5.

  (b) What will be the final value of the capacitor voltage when the charging is complete?

  (c) What will be the time constant of the discharge when the switch is opened?

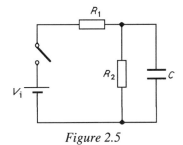

Figure 2.5

(d) Find the time constant of the charging of the capacitor in Figure 2.6, and the voltage across the capacitor one time constant after the closure of the switch.

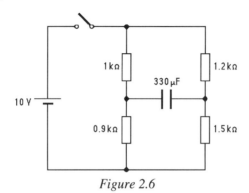

Figure 2.6

2.9 (a) A coil having inductance 20 mH and negligible resistance is connected across a 20 V r.m.s., 1 kHz sinusoidal supply. What is the amplitude of the current flowing in the circuit?

(b) The same coil is connected in series with a 220 Ω resistor. What is the time constant of the circuit?

(c) This series combination of resistor and inductor is connected, via a switch, across a 30 V d.c. supply. What will be the voltage across the inductor 0.2 ms after the switch closure?

(d) What will be the current flowing in the circuit 0.1 ms after the switch closure?

2.10 A transformer has a primary coil having 3000 turns and two secondary coils, coil 1 having 250 turns, coil 2 having 400 turns. The primary is connected to a 240 V, 50 Hz a.c. supply having negligible source resistance. The magnetising current of the transformer can be considered negligible.

(a) A 20 Ω resistor is connected across coil 1, while coil 2 is left open circuit. Calculate the voltage across the 20 Ω resistor, the current flowing in it and the current taken from the source.

(b) The 20 Ω resistor is left connected to coil 1 and a 16 Ω resistor is connected across coil 2. What will be the input resistance of the primary of the transformer, and what is the total power taken from the source?

SOLUTIONS

Question 2.1

(a)

The mean value is the average measured over one half-cycle and is therefore simply V_a.

The mean square value is simply V_a^2 (since the square value remains constant at V_a^2), and so the r.m.s. value is also V_a.

(b)

When a rectifying moving coil meter measures a sinusoidal waveform, it actually measures the mean value, which for a waveform of amplitude V_a is $V_a \times 2/\pi$. If the meter is calibrated to indicate r.m.s., it then displays this value as $V_a/\sqrt{2}$. Thus a mean value of V volts would be indicated as $V/\sqrt{2}$ multiplied by $\pi/2$, i.e. as $V \times \pi/(2\sqrt{2})$ volts. Thus the indicated value will be $\pi/(2\sqrt{2}) = 1.11$ times the actual mean value measured.

When the waveform of Figure 2.1 is rectified, its value remains constant at V_a and so the mean value is V_a. This will be displayed by the meter as $1.11 V_a$. Since, for the square wave, the actual r.m.s. value is also V_a, the displayed value will be in error by 11%.

Question 2.2

(a)

The waveform is shown in Figure 2.7.

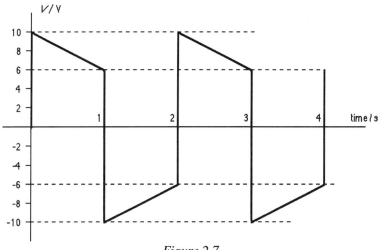

Figure 2.7

(b)

The mean value = $1/2 \times (10 \text{ V} + 6 \text{ V}) = 8$ V.

Signals, waveforms and a.c. components

The r.m.s. value can be calculated by first finding the mean square value.

The mean square value can be found by integrating over the first half cycle, since the symmetry means that the second half-cycle will have the same mean square value as the first. Hence,

$$\overline{v^2} = \frac{1}{T}\int_0^T v^2 \, dt$$

$$= \frac{1}{1}\int_0^1 (10 - 4t)^2 \, dt$$

$$= \int_0^1 \left(100 - 80t + 16t^2\right) dt$$

$$= \left[100t - 40t^2 + \frac{16t^3}{3}\right]_0^1$$

$$= 100 - 40 + \frac{16}{3}$$

$$= \frac{196}{3} \, V^2$$

So, the r.m.s. value is:

$$V_{r.m.s.} = \sqrt{(196/3)} \, V = 8.08 \, V.$$

Question 2.3

(a)

Summing all the sample values and dividing by the number of samples (23) gives the mean value as:

Mean = −12.5 V ÷ 23 = −0.54 V.

Squaring each value and summing (remember that all squared values will be positive), then dividing by the number of samples gives the mean-square value as 384.75 V² ÷ 23 = 16.73 V². The r.m.s. value can then be found by taking the square root of this value.

r.m.s. value = √16.73 V = 4.09 V.

(b)

One complete cycle is represented by 23 samples taken at 1 ms intervals. The period of the waveform is therefore 23 ms and the fundamental frequency is 1000/23 Hz = 43.5 Hz.

Question 2.4

(a)

In the expression $v = k_0 + k_1 \sin \omega_0 t + k_2 \sin 2\omega_0 t + k_3 \sin 3\omega_0 t + ...$, k_0 represents the average value of the waveform, which, from Figure 2.2 must be $V_a/2 = 5$ V.

Using the equation $k_n = V_a \left(\dfrac{\sin(n\pi/2)}{n\pi/2} \right)^2$

$k_1 = V_a \left(\dfrac{\sin \pi/2}{\pi/2} \right)^2 = V_a \left(\dfrac{1}{\pi/2} \right)^2 = V_a \left(\dfrac{2}{\pi} \right)^2 = 0.41 V_a = 4.1 \text{ V}$

$k_2 = V_a \left(\dfrac{\sin \pi}{\pi} \right)^2 = 0 \text{ V (since } \sin \pi = 0)$

$k_3 = V_a \left(\dfrac{\sin 3\pi/2}{3\pi/2} \right)^2 = V_a \left(\dfrac{2}{3\pi} \right)^2 = 0.045 V_a = 0.45 \text{ V}$

$k_4 = V_a \left(\dfrac{\sin 2\pi}{2\pi} \right)^2 = 0 \text{ V}$

$k_5 = V_a \left(\dfrac{\sin 5\pi/2}{5\pi/2} \right)^2 = V_a \left(\dfrac{2}{5\pi} \right)^2 = 0.016 V_a = 0.16 \text{ V}$

$k_6 = 0 \text{ V}$

The fundamental frequency = 1 / (25 ms) = 40 Hz.

The line spectrum is shown in Figure 2.8

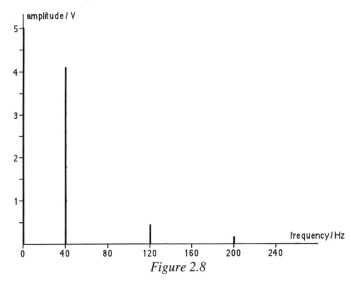

Figure 2.8

(b)

For a frequency component to have an amplitude less than 10 mV, the expression $V_a(2/n\pi)^2$ must have a value less than 0.01 V, and so the expression $(2/n\pi)^2$ must have a value less than 0.001.

Signals, waveforms and a.c. components

i.e.

$$\left(\frac{2}{n\pi}\right)^2 < 0.001$$

$$\frac{2}{n\pi} < \sqrt{0.001} = 0.0316$$

$$n > \frac{2}{0.0316\pi} = 20.15$$

Hence the twenty-first harmonic will be the first non-zero harmonic with an amplitude less than 10 mV.

Question 2.5

Considering Figure 2.3,

from 500 Hz to 1.5 kHz,

$$\overline{v^2} = \text{area under triangle}$$
$$= \tfrac{1}{2} \times \text{base} \times \text{height}$$
$$= \tfrac{1}{2} \times 1 \text{ kHz} \times 0.7 \times 10^{-3} \text{ V}^2 \text{ Hz}^{-1}$$
$$= 0.35 \text{ V}^2$$

from 1.5 kHz to 3 kHz,

$$\overline{v^2} = 1.5 \text{ kHz} \times 0.7 \times 10^{-3} \text{ V}^2 \text{ Hz}^{-1}$$
$$= 1.05 \text{ V}^2$$

from 3 kHz to 6 kHz,

$$\overline{v^2} = \tfrac{1}{2} \times 3 \text{ kHz} \times 0.7 \times 10^{-3} \text{ V}^2 \text{ Hz}^{-1}$$
$$= 1.05 \text{ V}^2$$

so, overall

$$\overline{v^2} = (0.35 + 1.05 + 1.05) \text{ V}^2 = 2.45 \text{ V}^2$$

Hence, r.m.s. voltage = 1.57 V

Question 2.6

(a)

$$\varepsilon = \varepsilon_0 \varepsilon_r = 8.854 \text{ pF m}^{-1} \times 350 \approx 3100 \text{ pF m}^{-1}$$

$$C = \frac{\varepsilon A}{d} = \frac{3100 \text{ pF m}^{-1} \times 20 \times 10^{-3} \text{ m} \times 250 \times 10^{-3} \text{ m}}{15 \times 10^{-6} \text{ m}}$$

$$= 1.03 \text{ μF}$$

(b)

At 2 kHz, the capacitive reactance is given by:

$$X_C = 1/\omega C = 1/2\pi f C = 1/\left(2\pi \times 2 \times 10^3 \text{ Hz} \times 1.03 \times 10^{-6} \text{ F}\right)$$
$$= 77 \text{ }\Omega$$

(c)

At 50 Hz, $\quad X_C = 77 \text{ }\Omega \times (2000 \text{ Hz})/(50 \text{ Hz}) = 3080 \text{ }\Omega$

Current $I = V/X_C = 240 \text{ V}/3080 \text{ }\Omega = 78 \text{ mA}$

Question 2.7

(a)

Time constant = $CR = 32 \times 10^{-6} \text{ F} \times 47 \times 10^3 \text{ }\Omega = 1.5 \text{ s}$

(b)

4.5 seconds is equal to $3CR$, after which time the capacitor voltage will have reached 95% of its final value, i.e. 14.25 V.

The output voltage will therefore be $(15 - 14.25)$ V = 0.75 V.

(c)

After 15 seconds, the value of the output voltage will be given by the equation:

$$V_{out} = V_i \exp\left(-t/CR\right)$$
$$= 15 \text{ V} \exp\left(-15/1.5\right) = 15 \text{ V} \exp(-10) = 0.7 \text{ mV}$$

The value of the capacitor voltage will therefore be 15 V minus 0.7 mV which is, for all practical purposes, 15 V.

(d)

The sketch of the variation of output voltage with time is shown in Figure 2.9.

When the switch is moved back to position 2, the capacitor voltage is 15 V, but the total e.m.f. in the circuit is zero. The initial discharge current flowing from the capacitor must therefore be such as to generate an initial voltage across the resistor of –15 V (so that $V_R + V_C = 0$ V). As the capacitor discharges, its voltage falls, so that the magnitude of the resistor voltage also falls. The result is that the output voltage initially drops practically instantaneously from zero to –15 V, then decays exponentially towards zero with time constant $CR = 1.5$ s, as shown in the figure.

Signals, waveforms and a.c. components

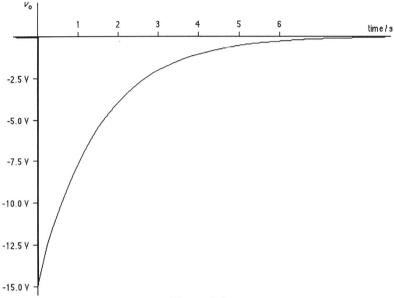

Figure 2.9

Question 2.8

(a)

Thévenin's theorem can be used to create the Thévenin equivalent of the circuit of Figure 2.5 with the switch closed and with the capacitor considered as the load.

Removing the capacitor, the component values of the Thévenin equivalent circuit are given by:

$$R_T = R_1 \text{ in parallel with } R_2 = \frac{R_1 R_2}{R_1 + R_2} \quad \text{and} \quad V_T = V_i \times \frac{R_2}{R_1 + R_2}$$

Hence, the time constant for charging the capacitor $= CR_T = \dfrac{CR_1 R_2}{R_1 + R_2}$

(b)

The final value of the capacitor voltage will be $V_T = \dfrac{V_i R_2}{R_1 + R_2}$

(c)

When the switch is opened, the capacitor can only discharge through R_2. The time constant of the discharge is therefore CR_2.

(d)

Again, a Thévenin equivalent circuit for Figure 2.6 can be created, with the capacitor considered as the load on the circuit.

Removing C gives Figure 2.10(a) and the equivalent circuit is Figure 2.10 (b).

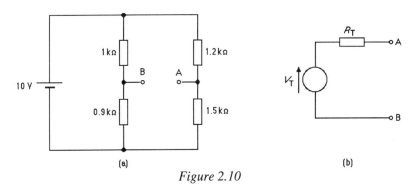

Figure 2.10

$V_T = V_A - V_B = 10\text{ V} \times 1.5/2.7 - 10\text{ V} \times 0.9/1.9 = 5.56\text{ V} - 4.74\text{ V} = 0.82\text{ V}$

Replacing the voltage source with a short-circuit gives:

$$R_T = \left(\frac{1.2 \times 1.5}{2.7} + \frac{1 \times 0.9}{1.9}\right)\text{k}\Omega = (0.667 + 0.474)\text{ k}\Omega = 1.14\text{ k}\Omega$$

The time constant is $= CR_T = 330 \times 10^{-9}\text{ F} \times 1.14 \times 10^3\text{ }\Omega = 376\text{ µs}$

The final capacitor voltage $= V_T = 0.82$ V, so the voltage across the capacitor after one time constant will be 63% of 0.82 V = 0.52 V

Question 2.9

(a)

The inductive reactance of the coil $= X_L = \omega L = 2\pi f L = 2\pi \times 10^3 \times 20 \times 10^{-3}\text{ }\Omega = 126\text{ }\Omega$

Therefore, r.m.s. current $= 20\text{ V} \div 126\text{ }\Omega = 159$ mA.

The amplitude of the current is therefore 159 mA $\times \sqrt{2}$ = 225 mA.

(b)

The time constant $= \dfrac{L}{R} = \dfrac{20 \times 10^{-3}\text{ H}}{220\text{ }\Omega} = 0.091$ ms

(c)

The inductor voltage after 0.2 ms is given by the equation:

$$v_L = V_S \exp(-tR/L) = V_S \exp(-0.2/0.091) = 30\text{ V} \times 0.111 = 3.33\text{ V}$$

(d)

After 0.1 ms, $\qquad v_L = V_S \exp(-0.1/0.091) = 30\text{ V} \times 0.333 = 10\text{ V}$

So, $\qquad v_R = V_S - v_L = 20\text{ V}$ and the current $i = 20\text{ V}/220\text{ }\Omega = 91$ mA

Signals, waveforms and a.c. components

Question 2.10

(a)

The voltage across the 20 Ω resistor is 240 V multiplied by the turns ratio.

i.e. $V = 240 \text{ V} \times 250/3000 = 20 \text{ V}$

The current flowing in the resistor is $I = 20 \text{ V} \div 20 \text{ Ω} = 1 \text{ A}$

The current taken from the source is equal to the secondary current multiplied by the turns ratio.

i.e.   Primary current $= 1 \text{ A} \times 250/3000 = 83 \text{ mA}$

(b)

The voltage across the 16 Ω resistor $= 240 \text{ V} \times 400/3000 = 32 \text{ V}$ and the current flowing in it is therefore 2 A.

Using the relationship $i_1 N_1 = i_2 N_2 + i_3 N_3$ (see answer to SAQ 21 on page 90 of the textbook), the current flowing in the primary can be calculated as:

$$i = 1 \text{ A} \times 250/3000 + 2 \text{ A} \times 400/3000 = 0.35 \text{ A}$$

The input resistance of the transformer is therefore 240 V ÷ 0.35 A = 686 Ω

The power taken from the source is $V \times I = 240 \text{ V} \times 0.35 \text{ A} = 84 \text{ W}$.

3 PHASOR ANALYSIS OF A.C. CIRCUITS

QUESTIONS

3.1 (a) Referring to the descriptions of capacitive and inductive reactance given in Section 2.5 of the textbook, and to the description of phasor diagrams given in Section 3.1 of the textbook draw, as accurately as you can, the phasor diagram showing the relationship between the current I flowing in the circuit of Figure 3.1 and the voltages V_R, V_L and V_C.

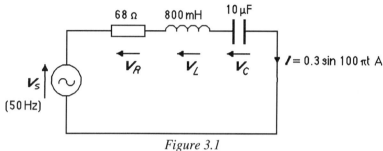

Figure 3.1

(b) On the same phasor diagram, construct the phasor representing the supply voltage, and hence deduce (i) the phase difference between the supply voltage and the current and (ii) the magnitude of the supply voltage.

(c) What relationship between V_L and V_C is necessary for the current to be in phase with the supply voltage?

3.2 (a) For each of the circuits in Figure 3.2, obtain an expression for the circuit impedance in $(a + jb)$ form.

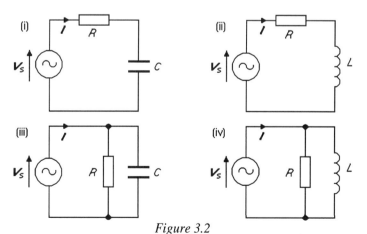

Figure 3.2

Phasor analysis of a.c. circuits

(b) If $R = 1.8$ kΩ, $C = 0.33$ µF and $L = 200$ mH, evaluate the expressions derived in (a) for $\omega = 3000$ rad s^{-1}.

(c) For each circuit, calculate the frequency at which the current taken from the source is 45° out of phase with the source voltage.

3.3 (a) For the circuit of Figure 3.3, show that the voltage transfer function $\mathbf{A} = V_o/V_s$ can be expressed in the form:

$$\mathbf{A} = \frac{R_2}{R_1 + R_2} \times \frac{1}{1 + j\omega T}$$

where $T = C\dfrac{R_1 R_2}{R_1 + R_2}$.

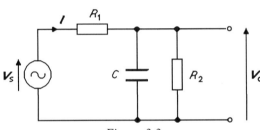

Figure 3.3

(b) Sketch the straight-line approximation to the Bode amplitude plot of the voltage transfer function when $R_1 = 1$ kΩ, $C = 100$ nF and $R_2 = 2$ kΩ.

3.4 (a) For the circuit of Figure 3.3, show that the impedance Z_{in} seen by the source can be expressed in the form:

$$\mathbf{Z}_{in} = (R_1 + R_2) \times \frac{1 + j\omega T_1}{1 + j\omega T_2}$$

where $T_1 = \dfrac{C R_1 R_2}{R_1 + R_2}$ and $T_2 = C R_2$

(b) For the same values of R_1, C and R_2 as in Question 3.3(b), sketch the variation of the magnitude of Z_{in} in the form of a straight-line approximation Bode plot, where the vertical axis is $20 \log_{10} |Z_{in}|$ dB.

3.5 (a) Show that the voltage transfer function of the circuit of Figure 3.4 can be expressed in the form

$$\frac{V_o}{V_s} = \frac{T_2}{T_1} \times \frac{(1 + j\omega T_1)}{(1 + j\omega T_2)}$$

where $T_1 = CR_1$ and $T_2 = \dfrac{CR_1 R_2}{R_1 + R_2}$

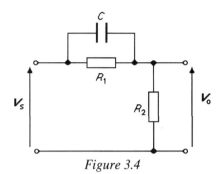

Figure 3.4

(b) If $R_1 = 4.7$ kΩ, $R_2 = 1.8$ kΩ and $C = 10$ μF, calculate the output voltage amplitude when $V_s = 10 \sin(50t)$ V. Also calculate the phase difference between the output voltage and the supply voltage.

(c) Sketch the straight-line approximation to the Bode amplitude plot of the voltage transfer function of the circuit, clearly labelling both axes.

3.6 (a) The voltage transfer function of the circuit of Figure 3.5 can be shown to be

$$\frac{V_O}{V_S} = \frac{j\omega(T_2 - T_1)}{(1 - \omega^2 T_1 T_2) + j\omega(T_1 + T_2)}$$

where $T_2 = CR_2$ and $T_1 = L/R_1$

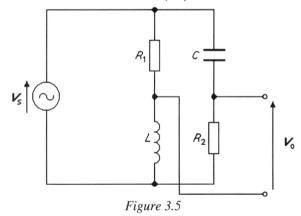

Figure 3.5

(i) What conditions will cause V_o to be zero when V_s is non-zero?

(ii) If V_o is not zero, at what frequency will the output voltage be in phase with the supply voltage?

(b) If $R_1 = R_2 = 1 \text{ k}\Omega$, $C = 2$ nF and $L = 1$ mH,

    (i) At what frequency is the output in phase with the supply?

    (ii) What is the amplitude of the output voltage when the supply voltage has amplitude 10 V and frequency 10 kHz, and what is its phase relative to the supply?

    (iii) Show that the angular frequency at which the output voltage leads the input voltage by 45° is given by the positive root of the quadratic equation

$$\omega^2 T_1 T_2 + \omega(T_1 + T_2) - 1 = 0$$

and evaluate this angular frequency for the component values given.

3.7 For the circuit of Figure 3.5 with the component and supply values given in Question 3.6 part (b), use Thévenin's theorem to find the current which flows in a load resistance of 2 kΩ connected across the output terminals, when the supply frequency is 10 kHz.

3.8 (a) Obtain an expression for the voltage transfer function of the network shown in Figure 3.6 in the form $(1 + j\omega T_1)/(1 + j\omega T_2)$.

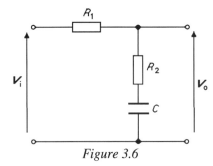

Figure 3.6

(b) Sketch the straight-line approximation to the Bode amplitude plot for the network when $R_1 = 7.5 \text{ k}\Omega$, $R_2 = 2.7 \text{ k}\Omega$ and $C = 0.1$ μF.

(c) Calculate the phase shift of the network at a frequency of 300 Hz.

(d) A buffer amplifier having a gain of 10 and zero phase shift over the range of frequencies of interest is placed between the output of the network of Figure 3.6 and the input of the network of Question 3.5 (Figure 3.4). Sketch the straight-line approximation to the Bode gain plot of the whole network, clearly labelling both axes.

(e) Draw a rough sketch of the Bode phase plot of the complete network.

Questions

3.9 An amplifier has the straight-line approximation Bode gain plot of Figure 3.7. It is placed between two identical single-lag CR networks with $C = 50$ nF and $R = 1$ kΩ.

(a) What is the voltage transfer function of the amplifier?

(b) What is the voltage transfer function of the complete network?

(c) Construct the straight-line approximation Bode gain and phase plots for the complete network.

(d) Calculate the total phase shift of the network at a frequency of 3 kHz.

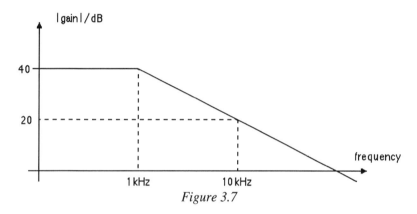

Figure 3.7

3.10 A coil having an inductance L and a resistance r is connected in parallel with a capacitor C across a current source I as shown in Figure 3.8.

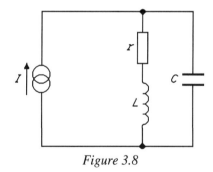

Figure 3.8

(a) Obtain an expression for the input admittance $\mathbf{Y}_T$ of the circuit in $(a + jb)$ form.

(b) If resonance is defined as the condition when $\mathbf{Y}_T$ becomes a pure conductance, show that the resonant angular frequency is given by:

$$\omega_0 = \sqrt{\frac{1}{LC} - \frac{r^2}{L^2}}$$

(c) Show that the dynamic impedance has the value L/Cr

(d) If $L = 1$ mH, $r = 10\ \Omega$ and $C = 1$ nF, calculate the resonant frequency and the dynamic impedance.

(e) If $|I| = 1$ mA, what is the amplitude of the voltage across the coil and the capacitor (i) at resonance and (ii) at very low frequency?

(f) What percentage error is introduced into the calculation of the resonant angular frequency if the expression for ω_0 is assumed to be $\omega_0 = 1/\sqrt{(LC)}$ instead of the correct expression given in (b)?

SOLUTIONS

Question 3.1

(a)

The resistor voltage is given by $|V_R| = |I| \times R$ and is in phase with the current.

The capacitor voltage is given by $|V_C| = |I| \times \frac{1}{\omega C}$ and lags the current by 90°.

The inductor voltage is given by $|V_L| = |I| \times \omega L$ and leads the current by 90°

Therefore,
$$|V_R| = |I| \times R = 0.3 \text{ A} \times 68 \text{ }\Omega = 20.4 \text{ V}$$

$$|V_C| = |I| \times \frac{1}{\omega C} = 0.3 \text{ A} \times \frac{1}{100\pi \times 10 \times 10^{-6}} \text{ }\Omega = 95.5 \text{ V}$$

$$|V_L| = |I| \times \omega L = 0.3 \text{ A} \times (100\pi \times 0.8) \text{ }\Omega = 75.4 \text{ V}$$

The phasor diagram is shown in Figure 3.9

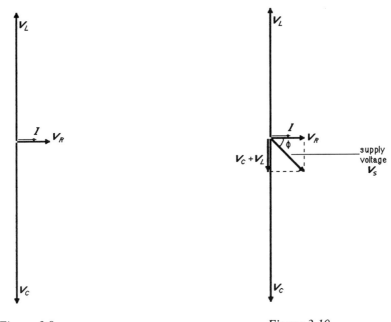

Figure 3.9　　　　　Figure 3.10

(b)

Since V_L and V_C are in opposite directions, the sum of the two phasors will be a phasor in the same direction as the larger, and with length equal to the difference in the lengths of the phasors, as shown in Figure 3.10.

Phasor analysis of a.c. circuits

(i) The supply voltage is equal to the phasor sum of V_R and $V_L + V_C$, as shown in the figure. The supply voltage therefore lags the current by an angle ϕ where:

$$\tan\phi = \frac{V_C + V_L}{V_R} = \frac{95.5 - 75.4}{20.4} = 0.985$$

Hence $\phi = \tan^{-1} 0.985 = 44.6°$

(ii) The magnitude of the supply voltage V is found by phasor addition and is:

$$|V| = \sqrt{|V_R|^2 + |V_C + V_L|^2} = \sqrt{20.4^2 + 20.1^2} = 28.6 \text{ V}$$

(c)

For the current to be in phase with the supply voltage, the length of the phasor representing $V_L + V_C$ must be zero, since that is the only condition which results in the phasor representing $V_R + V_L + V_C$ being in the same direction as the current I

Hence the magnitudes of V_L and V_C must be equal.

Question 3.2

(a)

(i) $Z = R + 1/j\omega C = R - j \cdot 1/\omega C$

(ii) $Z = R + j\omega L$

(iii) $Z = \dfrac{R \cdot 1/j\omega C}{R + 1/j\omega C} = \dfrac{R}{j\omega CR + 1} = \dfrac{R(1 - j\omega CR)}{1 + \omega^2 C^2 R^2} = \dfrac{R}{1 + \omega^2 C^2 R^2} - \dfrac{j\omega CR^2}{1 + \omega^2 C^2 R^2}$

(iv) $Z = \dfrac{R \cdot j\omega L}{R + j\omega L} = \dfrac{j\omega L R (R - j\omega L)}{R^2 + \omega^2 L^2} = \dfrac{\omega^2 L^2 R}{R^2 + \omega^2 L^2} + \dfrac{j\omega L R^2}{R^2 + \omega^2 L^2}$

(b)

(i) $Z = 1.8 \times 10^3 - j\dfrac{1}{3 \times 10^3 \times 0.33 \times 10^{-6}}$ Ω $= 1.8 \times 10^3 - j\dfrac{10^3}{0.99}$ Ω $= (1.8 - j)$ kΩ

(ii) $Z = 1.8 \times 10^3 + j\, 3 \times 10^3 \times 200 \times 10^{-3}$ Ω $= 1.8 \times 10^3 + j\, 600$ Ω $= (1.8 + j\, 0.6)$ kΩ

(iii) $(1 + \omega^2 C^2 R^2) = 1 + 9 \times 10^6 \times (0.33)^2 \times 10^{-12} \times (1.8)^2 \times 10^6 = 1 + 3.18 = 4.18$

so, $Z = \dfrac{1.8 \times 10^3}{4.18} - j\dfrac{3 \times 10^3 \times 0.33 \times 10^{-6} \times (1.8)^2 \times 10^6}{4.18}$ Ω $= (0.43 - j\, 0.77)$ kΩ

(iv) $R^2 + \omega^2 L^2 = (1.8)^2 \times 10^6 + 9 \times 10^6 \times 4 \times 10^{-2} = 3.6 \times 10^6$ Ω^2

$$Z = \left(\frac{9 \times 10^6 \times 4 \times 10^{-2} \times 1.8 \times 10^3}{3.6 \times 10^6} + j\frac{3 \times 10^3 \times 0.2 \times (1.8)^2 \times 10^6}{3.6 \times 10^6} \right) \Omega = 180 + j\,540\ \Omega$$

(c)

The current will be at 45° to the source voltage when the impedance phasor has angle + or – 45°.

Using the results from part (a) and equating real and imaginary parts,

(i)
$$R = 1/\omega C \quad \text{so} \quad \omega = 1/CR \quad \text{and} \quad f = 1/(2\pi CR)$$

so
$$f = \frac{1}{2\pi \times 1.8 \times 10^3 \times 0.33 \times 10^{-6}}\ \text{Hz} = 270\ \text{Hz}$$

(ii)
$$R = \omega L \quad \text{so} \quad \omega = R/L \quad \text{and} \quad f = R/(2\pi L)$$

so
$$f = \frac{1.8 \times 10^3}{2\pi \times 0.2}\ \text{Hz} = 1.4\ \text{kHz}$$

(iii) $R = \omega CR^2$, so $\omega = 1/CR$ and $f = 270$ Hz (as in part (i)).

(iv) $\omega^2 L^2 R = \omega L R^2$ so $\omega = R/L$ and $f = 1.4$ kHz (as in part (ii)).

Question 3.3

(a)

There are a number of different methods of tackling this problem, I have chosen to show three different approaches so that you can see which results in the simplest mathematical manipulation.

Method 1 treats the circuit as a voltage divider having two impedances Z_1 and Z_2 as shown in Figure 3.11, where Z_1 is simply R_1 and Z_2 is the parallel combination of C and R_2.

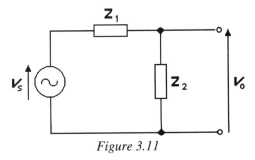

Figure 3.11

Phasor analysis of a.c. circuits

i.e. $\mathbf{Z_1} = R_1$ and $\mathbf{Z_2} = \dfrac{R_2 \cdot \frac{1}{j\omega C}}{R_2 + \frac{1}{j\omega C}} = \dfrac{R_2}{1 + j\omega C R_2}$

Using the voltage divider rule,

$$A = \dfrac{V_O}{V_S} = \dfrac{\frac{R_2}{(1+j\omega CR_2)}}{R_1 + \frac{R_2}{(1+j\omega CR_2)}} = \dfrac{R_2}{R_1 + j\omega CR_1 R_2 + R_2} = \dfrac{R_2}{(R_1 + R_2) + j\omega CR_1 R_2}$$

$$= \dfrac{R_2}{R_1 + R_2} \times \dfrac{1}{1 + j\omega CR_1 R_2 / (R_1 + R_2)}$$

Method 2 uses Thévenin's theorem to create a Thévenin equivalent circuit for V_S, R_1 and R_2, with C as the load on the circuit, as shown in Figure 3.12.

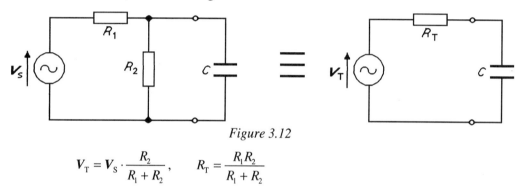

Figure 3.12

$$V_T = V_S \cdot \dfrac{R_2}{R_1 + R_2}, \qquad R_T = \dfrac{R_1 R_2}{R_1 + R_2}$$

By analogy with the result for the simple, low-pass CR circuit (p.113 of the textbook),

$$\dfrac{V_O}{V_T} = \dfrac{1}{1 + j\omega CR_T}$$

So, $\qquad \dfrac{V_O}{V_T} = \dfrac{1}{1 + j\omega CR_1 R_2 / (R_1 + R_2)}$

Hence,
$$\dfrac{V_O}{V_S} = \dfrac{V_T}{V_S} \times \dfrac{V_O}{V_T}$$

$$= \dfrac{R_2}{R_1 + R_2} \times \dfrac{1}{1 + j\omega CR_1 R_2 / (R_1 + R_2)}$$

Method 3 uses a Thévenin equivalent circuit with R_2 as the load, as shown in Figure 3.13.

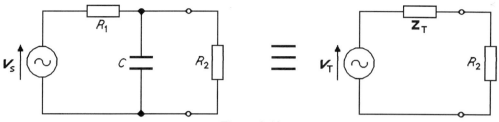

Figure 3.13

$$V_T = V_S \times \frac{1/j\omega C}{R_1 + 1/j\omega C} = V_S \times \frac{1}{1 + j\omega C R_1} \text{ and } Z_T = \frac{R_1 \cdot 1/j\omega C}{R_1 + 1/j\omega C} = \frac{R_1}{1 + j\omega C R_1}$$

$$\frac{V_O}{V_T} = \frac{R_2}{R_2 + Z_T} = \frac{R_2}{R_2 + R_1/(1 + j\omega C R_1)} = \frac{R_2(1 + j\omega C R_1)}{R_2 + j\omega C R_1 R_2 + R_1}$$

$$= \frac{R_2}{R_1 + R_2} \times \frac{1 + j\omega C R_1}{1 + j\omega C R_1 R_2/(R_1 + R_2)}$$

$$\frac{V_O}{V_S} = \frac{V_O}{V_T} \times \frac{V_T}{V_S} = \frac{V_O}{V_T} \times \frac{1}{(1 + j\omega C R_1)} = \frac{R_2}{R_1 + R_2} \times \frac{1}{1 + j\omega C R_1 R_2/(R_1 + R_2)}$$

(b)

When $R_1 = 1$ kΩ, $R_2 = 2$ kΩ and $C = 100$ nF,

$$\frac{R_2}{R_1 + R_2} = \frac{2 \times 10^3}{3 \times 10^3} = 0.67, \text{ so } \frac{R_1 R_2}{R_1 + R_2} = 0.67 \text{ kΩ}$$

Hence,

$$\frac{V_O}{V_S} = \frac{R_2}{R_1 + R_2} \times \frac{1}{1 + j\omega C R_1 R_2/(R_1 + R_2)}$$

$$= 0.67 \times \frac{1}{1 + j\omega \times 100 \times 10^{-9} \times 0.67 \times 10^3} = 0.67 \times \frac{1}{1 + j\omega \times 67 \times 10^{-6}}$$

The low frequency amplitude of the voltage transfer function = 0.67 (= −3.5 dB).

The corner frequency is at $\omega = 1/(67 \times 10^{-6})$ rad s^{-1} = 15×10^3 rad s^{-1}.

The straight line approximation Bode plot of the voltage transfer function is shown in Figure 3.14. Compare this with the generalised Bode plot shown in Figure 3.30 (page 120) of the text-book.

Phasor analysis of a.c. circuits

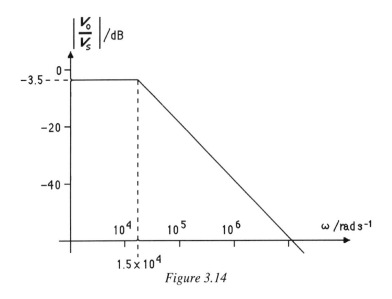

Figure 3.14

Question 3.4

(a)

The impedance $\mathbf{Z}_{in}$ seen by the source consists of R_1 in series with the parallel combination of C and R_2.

So,

$$\mathbf{Z}_{in} = R_1 + \frac{R_2 \cdot 1/j\omega C}{R_2 + 1/j\omega C} = R_1 + \frac{R_2}{j\omega C R_2 + 1}$$

$$= \frac{j\omega C R_1 R_2 + R_1 + R_2}{j\omega C R_2 + 1}$$

$$= (R_1 + R_2) \times \frac{1 + j\omega C R_1 R_2/(R_1 + R_2)}{1 + j\omega C R_2}$$

$$= (R_1 + R_2) \times \frac{1 + j\omega T_1}{1 + j\omega T_2}$$

(b)

When $R_1 = 1 \text{ k}\Omega$, $R_2 = 2 \text{ k}\Omega$ and $C = 100 \text{ nF}$,

$$T_1 = \frac{C R_1 R_2}{R_1 + R_2} = \frac{100 \times 10^{-9} \times 10^3 \times 2 \times 10^3}{3 \times 10^3} \text{ s} = 67 \times 10^{-6} \text{ s}$$

$$T_2 = C R_2 = 100 \times 10^{-9} \times 2 \times 10^3 \text{ s} = 200 \times 10^{-6} \text{ s}$$

So, $$\mathbf{Z}_{in} = 3 \times 10^3 \, \Omega \times \frac{1 + j\omega \times 67 \times 10^{-6}}{1 + j\omega \times 200 \times 10^{-6}}$$

At very low frequencies, when terms containing ω can be neglected, $\mathbf{Z}_{in} = 3 \times 10^3 \, \Omega$. This is in accordance with an intuitive examination of the circuit, where at low frequencies the reactance of the capacitor is very large, and the input impedance is simply the series combination of the two resistors.

At very high frequencies, when the 1's in the fraction can be neglected, $\mathbf{Z}_{in} = 10^3 \, \Omega$. This again accords with intuitive reasoning which says that at high enough frequencies the capacitor becomes a short circuit, so that the input impedance is simply R_1.

At intermediate frequencies, the way in which the impedance changes can be found by plotting the variation of $\mathbf{Z}_{in}$ with angular frequency. Figure 3.15 shows the straight line approximation to the plot of $20 \log_{10} |\mathbf{Z}_{in}|$ against angular frequency. The low-frequency value is $20 \log_{10} 3000 = 69.5$. The high-frequency value is $20 \log_{10} 1000 = 60$. The term $(1 + j\omega \times 200 \times 10^{-6})$ in the denominator introduces a corner from zero slope to a slope of -20 per decade at an angular frequency of $1/(200 \times 10^{-6})$ rad s^{-1} = 5000 rad s^{-1}. The term $(1 + j\omega \times 67 \times 10^{-6})$ in the numerator introduces a corner from a slope of -20 per decade to zero slope at a frequency of $1/(67 \times 10^{-6})$ rad s^{-1} = 15 000 rad s^{-1}.

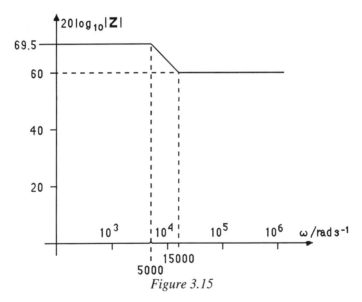

Figure 3.15

Question 3.5

(a)

Considering the circuit of Figure 3.4 to be made up of two impedances $\mathbf{Z}_1$ and $\mathbf{Z}_2$ as in Figure 3.11, where $\mathbf{Z}_2$ is simply R_2 and $\mathbf{Z}_1$ is the parallel combination of R_1 and C,

we can write $\quad \mathbf{Z}_2 = R_2$

$$Z_1 = \frac{R_1 \times 1/j\omega C}{R_1 + 1/j\omega C} = \frac{R_1}{1+j\omega CR_1}$$

$$\frac{V_O}{V_S} = \frac{Z_2}{Z_1+Z_2} = \frac{R_2}{R_1/(1+j\omega CR_1) + R_2}$$

$$= \frac{R_2(1+j\omega CR_1)}{R_1 + R_2 + j\omega CR_1 R_2}$$

$$= \frac{R_2}{R_1+R_2} \times \frac{(1+j\omega CR_1)}{\left(1 + j\omega CR_1 R_2/(R_1+R_2)\right)}$$

$$= \frac{T_2}{T_1} \times \frac{(1+j\omega T_1)}{(1+j\omega T_2)}$$

(b)

When $R_1 = 4.7$ kΩ, $R_2 = 1.8$ kΩ and $C = 10$ µF,

$$T_2 = \frac{CR_1 R_2}{R_1+R_2} = \frac{10 \times 10^{-6} \times 4.7 \times 10^3 \times 1.8 \times 10^3}{6.5 \times 10^3} = 1.3 \times 10^{-2} \text{ s}$$

$$T_1 = CR_1 = 10 \times 10^{-6} \times 4.7 \times 10^3 = 4.7 \times 10^{-2} \text{ s}$$

$$\frac{V_O}{V_S} = \frac{T_2}{T_1} \times \frac{(1+j\omega T_1)}{(1+j\omega T_2)} = \frac{1.3}{4.7} \times \frac{\left(1+j\omega \times 4.7 \times 10^{-2}\right)}{\left(1+j\omega \times 1.3 \times 10^{-2}\right)}$$

When $V_S = 10 \sin(50t)$, $|V_S| = 10$ V and $\omega = 50$ rad s^{-1}, so:

$$V_O = 10 \times 0.28 \times \frac{(1+j2.35)}{(1+j0.65)} \text{ V}$$

$$|V_O| = 2.8 \times \frac{\sqrt{1+2.35^2}}{\sqrt{1+0.65^2}} = 6.0 \text{ V}$$

$$\angle V_O = \tan^{-1} 2.35 - \tan^{-1} 0.65 = 34°$$

(c)

The voltage transfer function is $\dfrac{V_O}{V_S} = \dfrac{T_2}{T_1} \times \dfrac{(1+j\omega T_1)}{(1+j\omega T_2)} = 0.28 \times \dfrac{\left(1+j\omega \times 4.7 \times 10^{-2}\right)}{\left(1+j\omega \times 1.3 \times 10^{-2}\right)}$

The numerator term will provide an upward break point at an angular frequency given by

$$\omega = 1/(4.7 \times 10^{-2}) = 21.2 \text{ rad s}^{-1}$$

The denominator term will provide a downward break point at an angular frequency given by

$$\omega = 1/(1.3 \times 10^{-2}) = 77 \text{ rad s}^{-1}$$

The low frequency gain is 0.28 (or −11 dB), while the high frequency gain is 1 (or 0 dB).

The required straight-line approximation to the Bode gain plot of the voltage transfer function is shown in Figure 3.16.

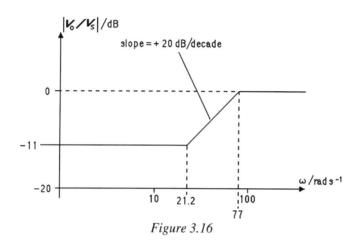

Figure 3.16

Question 3.6

(a)

(i) V_o will be zero when $T_2 = T_1$, so making the numerator of the voltage transfer function zero. Under this condition, V_o will be zero whatever the frequency of the supply voltage.

V_o will also be very small at very low frequencies (ω approaching zero) when the numerator becomes much smaller than the denominator, and at very high frequencies (ω approaching infinity) when the ω^2 term in the denominator becomes much larger than the ω term in the numerator.

(ii) If V_o is not zero, the output voltage will be in phase with the supply voltage when $(1 - \omega^2 T_1 T_2)$ is zero. In this case, the real part of both the numerator and denominator will be zero, so that both numerator and denominator have a phase angle of 90°. The total phase angle of the voltage transfer ratio is then zero. Hence the frequency at which the output voltage is in phase with the supply voltage is given by

$$\omega = 1/\sqrt{T_1 T_2} \quad \text{or} \quad f = 1/(2\pi\sqrt{T_1 T_2})$$

Phasor analysis of a.c. circuits

(b)

(i) Using the values $R_1 = R_2 = 1$ kΩ, $C = 2$ nF and $L = 1$ mH,

$$T_1 = L/R_1 = 10^{-3}/10^3 = 10^{-6} \text{ s}$$

$$T_2 = CR_2 = 2 \times 10^{-9} \times 10^3 = 2 \times 10^{-6} \text{ s}$$

The frequency at which the output voltage is in phase with the supply is given by

$$f = 1/(2\pi\sqrt{T_1 T_2}) = 1/(2\pi\sqrt{2 \times 10^{-12}}) \text{ Hz} = 10^6/(2\pi\sqrt{2}) \text{ Hz}$$

$$= 112.5 \text{ kHz}$$

(ii) Substituting the values for T_1 and T_2, together with $\omega = 2\pi \times 10^4$ rad s^{-1} (i.e. $f = 10$ kHz) into the expression for the voltage transfer function gives:

$$\frac{V_O}{V_S} = \frac{j\omega(T_2 - T_1)}{(1 - \omega^2 T_1 T_2) + j\omega(T_1 + T_2)}$$

$$= \frac{j \times 2\pi \times 10^4 \times 10^{-6}}{(1 - 4\pi^2 \times 10^8 \times 2 \times 10^{-12}) + j \times 2\pi \times 10^4 \times 3 \times 10^{-6}}$$

$$= \frac{j \times 2\pi \times 10^{-2}}{(1 - 8\pi^2 \times 10^{-4}) + j \times 6\pi \times 10^{-2}}$$

$$= \frac{j\, 0.063}{0.992 + j\, 0.188}$$

Hence,

$$V_o = \frac{j\, 0.63}{0.992 + j\, 0.188}$$

$$|V_o| = \frac{0.63}{\sqrt{0.992^2 + 0.188^2}} = 0.62 \text{ V}$$

$$\angle V_o = -\tan^{-1}(0.188/0.992) = -10.7°$$

(iii) The output will lead the input voltage by 45° when the phase of the denominator of the voltage transfer function is 45° (since the phase of the numerator is 90°).

This occurs when the real and imaginary parts of the denominator are equal, i.e when:

$$1 - \omega^2 T_1 T_2 = \omega(T_1 + T_2)$$

or

$$\omega^2 T_1 T_2 + \omega(T_1 + T_2) - 1 = 0$$

The roots of this equation are:

$$\omega = \frac{-(T_1+T_2) \pm \sqrt{(T_1+T_2)^2 - 4 \times T_1 T_2 \times (-1)}}{2T_1 T_2}$$

$$= \frac{-(T_1+T_2) \pm \sqrt{(T_1+T_2)^2 + 4T_1 T_2}}{2T_1 T_2}$$

Clearly, since the quantity under the square root is greater than $(T_1+T_2)^2$, one root is negative and one is positive. Since ω must be a positive quantity, the required value must be equal to the positive root.

For the component values given,

$$\omega = \frac{-3 \times 10^{-6} + \sqrt{9 \times 10^{-12} + 8 \times 10^{-12}}}{4 \times 10^{-12}}$$

$$= \frac{-3 + \sqrt{17}}{4 \times 10^{-6}} = 0.28 \times 10^6 \quad \text{rad s}^{-1}.$$

Question 3.7

Figure 3.17 shows the circuit of Figure 3.5 with the load connected, and this is re-drawn in Figure 3.18 with the load removed and the voltage source replaced by its zero internal impedance.

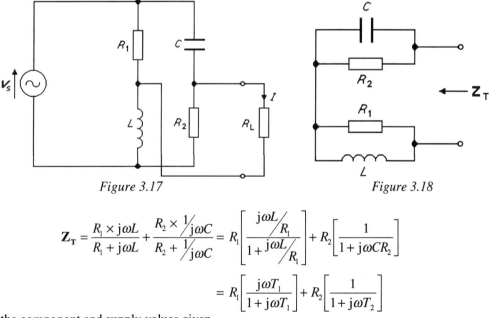

Figure 3.17   *Figure 3.18*

$$Z_T = \frac{R_1 \times j\omega L}{R_1 + j\omega L} + \frac{R_2 \times \frac{1}{j\omega C}}{R_2 + \frac{1}{j\omega C}} = R_1 \left[\frac{j\omega L / R_1}{1 + j\omega L / R_1} \right] + R_2 \left[\frac{1}{1 + j\omega C R_2} \right]$$

$$= R_1 \left[\frac{j\omega T_1}{1 + j\omega T_1} \right] + R_2 \left[\frac{1}{1 + j\omega T_2} \right]$$

For the component and supply values given,

Phasor analysis of a.c. circuits

$$\omega T_1 = \omega L / R = \frac{2\pi \times 10^4 \times 10^{-3}}{10^3} = 0.063$$

$$\omega T_2 = \omega C R_2 = 2\pi \times 10^4 \times 2 \times 10^{-9} \times 10^3 = 0.126$$

$$\mathbf{Z}_T = 10^3 \left[\frac{j0.063}{1+j0.063} \right] + 10^3 \left[\frac{1}{1+j0.126} \right]$$

$$= 10^3 \left[\frac{j0.063 + j^2 0.008 + 1 + j0.063}{(1+j0.063)(1+j0.126)} \right]$$

$$= 10^3 \left[\frac{0.992 + j0.126}{1 + j^2 0.008 + j0.188} \right]$$

$$= 10^3 \left[\frac{0.992 + j0.126}{0.992 + j0.188} \right]$$

The open-circuit output voltage V_T has already been calculated in Question 3.6, part (b) to be:

$$V_T = \frac{j0.63}{0.992 + j0.188}$$

The current through the 2 kΩ load resistor is therefore:

$$I = \frac{V_T}{Z_T + R_L} = \frac{j0.63 / (0.992 + j0.188)}{10^3 \left[\frac{0.992 + j0.126}{0.992 + j0.188} \right] + 2 \times 10^3} \text{ A}$$

$$= \frac{j0.63}{0.992 + j0.126 + 1.984 + j0.376} \text{ mA}$$

$$= \frac{j0.63}{2.976 + j0.502} \text{ mA}$$

$$|I| = \frac{0.63}{\sqrt{2.976^2 + 0.502^2}} \text{ mA} = \frac{0.63}{3.02} \text{ mA} = 0.21 \text{ mA}$$

$$\angle I = 90° - \tan^{-1} \frac{0.502}{2.976} = 80.4°$$

Hence, $\quad I = 0.21 \angle 80.4° \text{ mA}$

Question 3.8

(a)

$$\frac{V_o}{V_i} = \frac{R_2 + 1/j\omega C}{R_1 + R_2 + 1/j\omega C} = \frac{1+j\omega CR_2}{1+j\omega C(R_1+R_2)} = \frac{1+j\omega T_1}{1+j\omega T_2}$$

where $T_1 = CR_2$ and $T_2 = C(R_1+R_2)$

(b)

When $R_1 = 7.5\ \text{k}\Omega$, $R_2 = 2.7\ \text{k}\Omega$ and $C = 0.1\ \mu\text{F}$,

$$\frac{V_o}{V_i} = \frac{1+j\omega \times 0.27 \times 10^{-3}}{1+j\omega \times 1.02 \times 10^{-3}}$$

The numerator provides an upward break point at

$$\omega = \frac{1}{0.27 \times 10^{-3}}\ \text{rad s}^{-1} = 3.7 \times 10^3\ \text{rad s}^{-1}$$

The denominator provides a downward break point at

$$\omega = \frac{1}{1.02 \times 10^{-3}}\ \text{rad s}^{-1} = 0.98 \times 10^3\ \text{rad s}^{-1}$$

The low frequency gain (as ω approaches zero) = 1 (= 0 dB).

The high frequency gain = $0.27 \div 1.02 = 0.26$ (= -11.5 dB).

The straight-line approximation to the Bode gain plot is shown in Figure 3.19

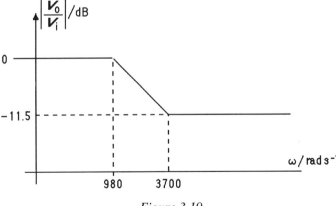

Figure 3.19

Phasor analysis of a.c. circuits

(c)

At 300 Hz, $\omega = 2\pi f = 1.9 \times 10^{-3}$ rad s^{-1}, so

$$\angle \frac{V_o}{V_i} = \tan^{-1} \omega T_1 - \tan^{-1} \omega T_2 = \tan^{-1}(1.9 \times 10^3 \times 0.27 \times 10^{-3}) - \tan^{-1}(1.9 \times 10^3 \times 1.02 \times 10^{-3})$$

$$= 27.2° - 62.7° = -35.5°$$

(d)

The Bode gain plot will be the combination of the plots of Figures 3.19 and 3.16, plus a frequency-independent gain of 20 dB from the amplifier. This combination is shown in Figure 3.20.

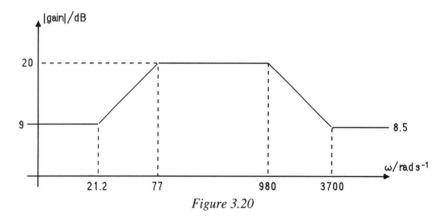

Figure 3.20

(e)

The Bode phase plot is sketched in Figure 3.21.

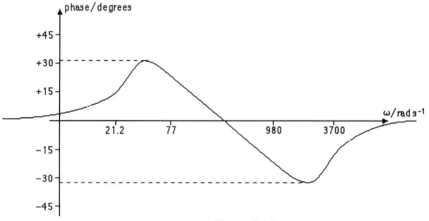

Figure 3.21

The phase is zero at both very low and very high frequencies (corresponding with the horizontal portions of the Bode gain plot), is positive when the gain plot has an upward slope and is negative when the gain plot has a downward slope. You have already calculated the phase of one network at an angular frequency of 50 rad s^{-1} to be +33°, and at this low frequency the phase of the other network is very small, so that the phase of the combined network is almost +33°. You have also just calculated the phase of the network of Figure 3.6 at a frequency of 1.9×10^3 rad s^{-1} to be −35.5°, and at this frequency the phase of the other network is very small.

Figure 3.21 shows all these phase values and also shows the approximate shape of the curve joining these points.

Question 3.9

(a)

The low-frequency gain of the amplifier is +40 dB or 100.

The downward break-point is at a frequency of 1 kHz or 6.28×10^3 rad s^{-1}.

The voltage transfer function of the network is therefore:

$$A = \frac{100}{(1 + j\omega/\omega_c)} = \frac{100}{(1 + j\omega \times 1.6 \times 10^{-4})}$$

(b)

The voltage transfer function of a single-lag CR network having $C = 50$ nF and $R = 1$ kΩ is:

$$A = \frac{1}{(1 + j\omega \times 50 \times 10^{-9} \times 10^3)} = \frac{1}{(1 + j\omega \times 5 \times 10^{-5})}$$

The voltage transfer function of the complete network is therefore:

$$A = \frac{100}{(1 + j\omega \times 1.6 \times 10^{-4})(1 + j\omega \times 5 \times 10^{-5})(1 + j\omega \times 5 \times 10^{-5})}$$

(c)

The Bode gain plot of the complete network will have a low frequency value of 40 dB, will have a downward break at $f = 1$ kHz from the amplifier, and will have a double downward break point (i.e. from a slope of −20 dB/decade to a slope of −60 dB/decade) at $\omega = 1/(5 \times 10^{-5})$ rad s^{-1} $= 2 \times 10^4$ rad s^{-1}, i.e. $f = 10^4/\pi$ Hz $= 3.2$ kHz, from the two equal single-lag circuits.

Figure 3.22 shows the straight-line approximation Bode gain plot.

The straight-line approximate phase plot will be the combination of three phase plots like that of Figure 3.31 of the textbook, one with a break-point frequency of 1 kHz (6.28×10^3 rad s^{-1}), the other two with break-point frequency 3.2 kHz (2×10^4 rad s^{-1}). The combined phase sum of the three is shown in Figure 3.23.

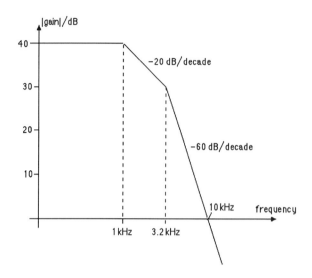

Figure 3.22

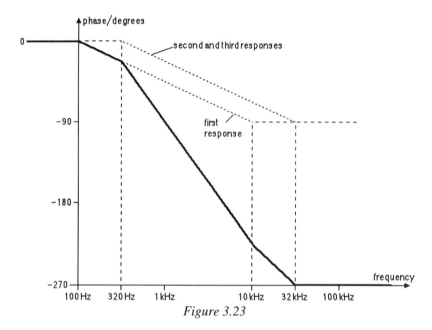

Figure 3.23

(d)

At a frequency of 3 kHz, $\omega = 2\pi \times 3 \times 10^3$ rad s^{-1} = 18.8×10^3 rad s^{-1} and the phase shift will be:

$$\phi = -\tan^{-1}\left(\frac{18.8 \times 10^3}{6.28 \times 10^3}\right) - 2 \times \tan^{-1}\left(\frac{18.8 \times 10^3}{2 \times 10^4}\right)$$
$$= -71.6° - 2 \times (43.3°) = -158°$$

Solutions

Question 3.10

(a)

$$Y_T = Y_{coil} + Y_{capacitor} = \frac{1}{r + j\omega L} + j\omega C$$

$$= \frac{r - j\omega L}{r^2 + \omega^2 L^2} + j\omega C$$

$$= \frac{r}{r^2 + \omega^2 L^2} + j\left(\omega C - \frac{\omega L}{r^2 + \omega^2 L^2}\right).$$

(b)

At resonance, since Y_T is a pure conductance, so $\left(\omega C - \frac{\omega L}{r^2 + \omega^2 L^2}\right) = 0$.

Hence,

$$C - \frac{L}{r^2 + \omega_0^2 L^2} = 0$$

$$C(r^2 + \omega_0^2 L^2) = L \quad (1)$$

$$\omega_0^2 L^2 C = L - Cr^2$$

$$\omega_0^2 = \frac{1}{LC} - \frac{r^2}{L^2}$$

$$\omega_0 = \sqrt{\frac{1}{LC} - \frac{r^2}{L^2}}.$$

(c)

At resonance, $Y_T = \frac{r}{r^2 + \omega_0^2 L^2}$ but, from equation (1) above, $r^2 + \omega_0^2 L^2 = L/C$ so that $Y_T = \frac{rC}{L}$ and the dynamic impedance is simply $L/(Cr)$.

(d)

When $L = 1$ mH, $r = 10\ \Omega$ and $C = 1$ nF,

$$\omega_0 = \sqrt{\frac{1}{10^{-3} \times 10^{-9}} - \frac{10^2}{10^{-6}}}\ \text{rad s}^{-1} = \sqrt{10^{12} - 10^8}\ \text{rad s}^{-1} \approx 10^6\ \text{rad s}^{-1}.$$

Hence, the resonant frequency = $\omega_0/2\pi$ = 160 kHz.

The dynamic impedance is given by: $Z_d = L/Cr = \dfrac{10^{-3}}{10^{-9} \times 10} = 10^5\ \Omega$

Phasor analysis of a.c. circuits

(e)

(i) If $|I| = 1$ mA, the amplitude of the voltage across both the coil and the capacitor at resonance is 10^{-3} A $\times$ 10^5 Ω = 100 V.

(ii) At very low frequency, the reactance of the coil is simply r because ωL will be very small, while the reactance of the capacitor is infinitely large. The circuit impedance is therefore r and the amplitude of the voltage across the circuit is 10^{-3} A $\times$ 10 Ω = 10 mV.

(f)

As can be seen from the calculation of ω_0 in part (d), the error introduced into the calculation of ω_0^2 by ignoring the second term in the expression is 10^8 in 10^{12} or 1 part in 10^4. The error in the calculation of ω is therefore half this amount or 5 parts in 10^5.

[**Note:** this halving of the error when taking a square root can be justified as follows:

If an actual quantity x_a is measured with a fractional error of ε, then the measured value x_m can be expressed as $x_m = x_a(1 \pm \varepsilon)$.

The error in the measurement of x_a^2 can then be evaluated as:

$$x_m^2 = x_a^2(1 \pm \varepsilon)^2 = x_a^2(1 \pm 2\varepsilon + \varepsilon^2)$$

If ε is very much smaller than 1, the ε^2 can be neglected and this expression becomes

$$x_m^2 = x_a^2(1 \pm 2\varepsilon)$$

so, if the error in the measurement of x_a is ε, the error in the measurement of x_a^2 is 2ε. Logically, therefore, it follows that if the error in the calculation of ω_0^2 is 1 part in 10^4, the resulting error in the calculation of ω_0 must be half of this value.]

4 AMPLIFIERS AND FEEDBACK

QUESTIONS

4.1 (a) In SAQ 2 of Chapter 4 of the textbook (Page 158), an amplifier with an open-circuit voltage gain A_v of 100 is shown to produce an 825-fold increase in the signal developed across a load by a given signal source. Explain why this occurs.

(b) In the situation described in SAQ 3 of the same chapter, if $R_s = R_L$, by what factor is the voltage across the load increased when the amplifier is introduced between the source and the load? Explain why this increase factor is so much less than that in SAQ 2.

4.2 An operational amplifier having the following characteristics:

Low-frequency gain A_v $= 2 \times 10^3$
Input resistance r_{in} $= 60 \text{ k}\Omega$
Output resistance r_{out} $= 2 \text{ k}\Omega$

is connected in the circuit configuration of Figure 4.1.

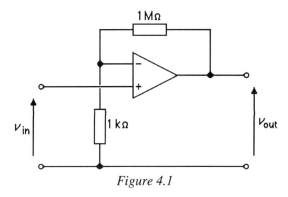

Figure 4.1

Calculate the low-frequency closed-loop gain, input resistance and output resistance of the feedback amplifier.

4.3 The amplifier of Question 4.2 is now replaced by a compensated operational amplifier having the following characteristics:

Open-circuit gain $\mathbf{A}_v$ $= 10^5 / (1 + j\,0.016\omega)$
Input impedance $\mathbf{z}_i$ $= 100 \text{ k}\Omega$
Output impedance $\mathbf{z}_o$ $= 100 \text{ }\Omega$

Calculate the closed-loop gain, input impedance and output impedance of the feedback amplifier at (a) 10 Hz and (b) 1 kHz.

Amplifiers and feedback

4.4 The amplifier in the circuit of Figure 4.2(a) has the open-loop gain/frequency characteristic shown in Figure 4.2(b). Assuming that the effects of the amplifier's input and output impedances can be ignored, calculate the magnitude of the closed-loop gain at frequencies of 15, 30, 60, 120 and 240 kHz. Plot these results on the axes provided in Figure 4.3 and read off the 3dB bandwidth. Compare the open-loop and closed-loop gain-bandwidth products.

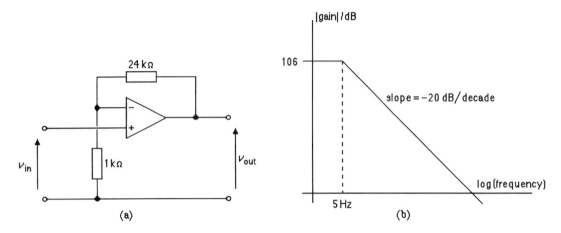

Figure 4.2

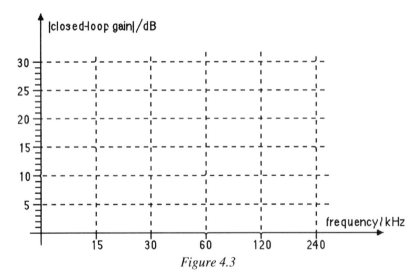

Figure 4.3

4.5 In the circuit of Figure 4.4, if the low-frequency differential gain A_V of the amplifier is sufficiently large, then v_i is a very small voltage which can be considered zero. Assume that the input impedance of the amplifier is large enough to be considered infinite, and the output impedance small enough to be ignored.

(a) Assuming that v_i is zero, show that, for the circuit of Figure 4.4, $V_o = \dfrac{R_2}{R_1}(V_2 - V_1)$

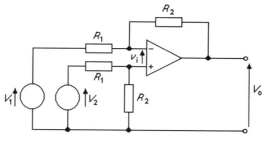

Figure 4.4

(b) For the circuit of Figure 4.5, show that $V_o = R_E\left(\dfrac{V_A}{R_A} + \dfrac{V_B}{R_B}\right)$ where R_E is the equivalent resistance of R_A, R_B and R_C in parallel.

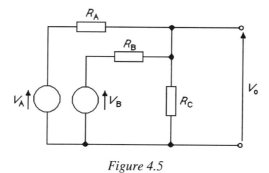

Figure 4.5

(c) Using the results of parts (a) and (b), evaluate V_o in terms of V_1, V_2 and V_3 in the circuit of Figure 4.6.

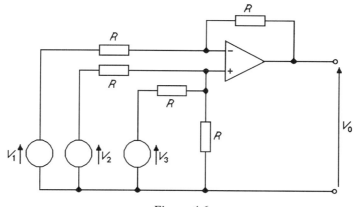

Figure 4.6

Amplifiers and feedback

4.6 In the feedback circuit of Figure 4.7, $Z_S = 2(1 - j\,0.01\omega)$ kΩ and $Z_F = 20$ kΩ. Assuming that the virtual earth approximation is valid, what will be the closed-loop gain at angular frequencies of 10, 100 and 1000 rad s^{-1}?

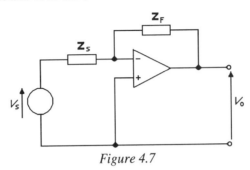

Figure 4.7

4.7 **(a)** By calculating the Thévenin equivalent circuit parameters for the input circuit of the feedback amplifier of Figure 4.8, show that the closed-loop gain can be expressed in the form

$$G = V_o/V_s = \frac{R_F}{R_1 + R_2} \times \frac{1}{1 + j\omega T} \quad \text{where} \quad T = \frac{CR_1 R_2}{R_1 + R_2}$$

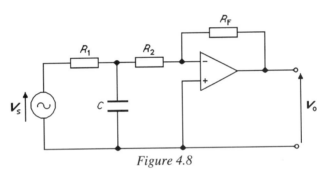

Figure 4.8

(b) If, in Figure 4.8, $R_1 = R_2 = 5$ kΩ, $R_F = 200$ kΩ and $C = 0.5$ µF, sketch the closed-loop Bode gain plot over the frequency range 1 Hz to 10 kHz. (Assume that the amplifier has adequate gain over this frequency range for the virtual earth approximation to be valid.)

(c) If the amplifier has gain $A_v = 2 \times 10^5/(1 + j\,0.03\omega)$, is the virtual earth assumption valid over the range of frequencies of interest? Explain your answer.

4.8 An operational amplifier has the following offset characteristics:

Input offset voltage V_{IO}	= 2 mV
Input bias current I_B	= 150 nA
Input offset current I_{IO}	= 50 nA

(a) The amplifier is used in the circuit of Figure 4.9. Calculate the output offset voltage.

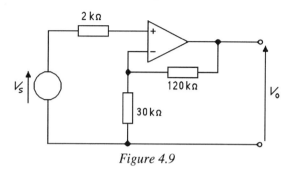

Figure 4.9

(b) What simple change to the circuit will minimise the contribution of I_B to the output offset voltage without changing any other circuit characteristic?

4.9 A 741 operational amplifier having the noise characteristics shown in Figure 4.31 of the textbook (page 190) is connected into the circuit shown in Figure 4.10.

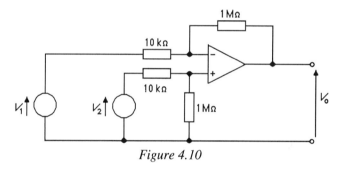

Figure 4.10

(a) Calculate the output noise voltage of the amplifier in the frequency band 10 Hz to 100 kHz, assuming negligible noise in the other circuit components.

(b) V_1 and V_2 are similar signals, differing only slightly in amplitude, so that V_2 can be expressed as $V_1 + v$ where v is a small amplitude signal at the same frequency as V_1. Assuming that the input signal frequency falls within the frequency band 10 Hz to 100 kHz, and that all other frequencies can be filtered out of the output signal, what is the minimum value of v which will provide a signal-to-noise ratio of at least 1000?

4.10 The amplifier in the circuit of Figure 4.9 has a slew rate of 10 V μs^{-1}. The input signal is a sinewave voltage of amplitude 2 V.

(a) What is the maximum signal frequency which can be amplified without distortion due to slew rate?

(b) Assuming that the amplifier has adequate gain-bandwidth product, what is the maximum input signal amplitude at a frequency of 1 MHz which can be amplified without distortion caused by slew rate?

Amplifiers and feedback

SOLUTIONS

Question 4.1

(a)

The amplifier performs two functions, firstly it amplifies the input voltage and secondly it reduces the loading on the source caused by the load.

In this case, without the amplifier, the signal voltage from the source, when connected directly to the load, is attenuated by a factor of 10 (since the source resistance is 9 kΩ and the load resistance is 1 kΩ). With the amplifier interposed between the source and the load, the amplifier itself loads the source, but by much less, and the attenuation of the signal by the input resistance of the amplifier is by a factor of 1.1 only (since the amplifier input resistance is 91 kΩ). At the output of the amplifier, the attenuation of the signal caused by the load on the amplifier output resistance is again a factor of 1.1 only (since the load resistance is 1 kΩ and the amplifier output resistance is 100 Ω). Combining these two attenuations with the amplifier gain of 100 gives an overall voltage gain of about 82.5 compared with the attenuation of 10 which occurs without the amplifier.

(b)

In SAQ 3, if $R_s = R_L$ then, without the amplifier, the signal would be attenuated by a factor of 2 because the load on the source would equal the source resistance. When the amplifier is interposed, because $R_s = r_{in} = r_{out} = R_L$, there will be two attenuation factors of 2, combined with a gain of 10, giving an overall gain of 2.5 and an increase in load voltage by a factor of 5.

This is much less than the increase factor of (a) because (i) the amplifier gain is lower, (ii) the amplifier input loads the source more and (iii) the load causes greater attenuation at the amplifier output because of the higher output resistance of the amplifier relative to the load resistance.

Question 4.2

Low frequency gain

$$\frac{v_{out}}{v_{in}} = \frac{A_v}{(1+\beta A_v)} = \frac{2 \times 10^3}{1 + 10^{-3} \times 2 \times 10^3} = 667$$

Input resistance

$$R_i = (1+\beta A_v)r_{in} = 3 \times 60 \text{ k}\Omega = 180 \text{ k}\Omega$$

Output resistance

$$R_o = \frac{r_o}{(1+\beta A_v)} = \frac{2}{3} \text{ k}\Omega = 667 \text{ }\Omega$$

Do not make the mistake of assuming that βA_v is much greater than 1 without checking. In this case its value is only 2.

Question 4.3

With the new amplifier,

$$G = \frac{A_v}{1+\beta A_v} = \frac{10^5/(1+j0.016\omega)}{1+10^{-3} \times 10^5/(1+j0.016\omega)} = \frac{10^5}{101+j0.016\omega}$$

$$Z_i = (1+\beta A_v)z_i = \left(1+\frac{10^{-3} \times 10^5}{1+j0.016\omega}\right) \times 10^5 \ \Omega$$

$$= \frac{(101+j0.016\omega)}{(1+j0.016\omega)} \times 10^5 \ \Omega$$

$$Z_o = z_o/(1+\beta A_v) = \frac{100}{\left(1+10^2/(1+j0.016\omega)\right)} \ \Omega$$

$$= \frac{(1+j0.016\omega)}{(101+j0.016\omega)} \times 100 \ \Omega$$

(a)

At 10 Hz, $\omega = 20\pi$ rad s^{-1} and $0.016\omega \approx 1$.

Hence, $\quad G = \dfrac{10^5}{101+j} = 990\angle -0.6°$

$$Z_i = \frac{(101+j)}{(1+j)} \times 10^5 \ \Omega = 7.1\angle -44.4° \ M\Omega$$

$$Z_o = \frac{(1+j)}{(101+j)} \times 100 \ \Omega = 1.4\angle 44.4° \ \Omega$$

(b)

At 1 kHz, $\omega = 2000\pi$ rad s^{-1} and $0.016\omega \approx 100$.

Hence, $\quad G = \dfrac{10^5}{101+j100} = 704\angle -44.7°$

$$Z_i = \frac{(101+j100)}{(1+j100)} \times 10^5 \ \Omega = 0.14\angle -44.7° \ M\Omega$$

$$Z_o = \frac{(1+j100)}{(101+j100)} \times 100 \ \Omega = 70\angle 44.7° \ \Omega$$

Amplifiers and feedback

Question 4.4

From the graph of gain amplitude against frequency, the low-frequency gain is 106 dB which is a gain of 2×10^5, while the break-point is at 5 Hz.

The open-loop gain is therefore $A_v = \dfrac{2 \times 10^5}{1 + jf/5} = \dfrac{2 \times 10^5}{1 + j0.2f}$

The closed-loop gain $G = \dfrac{A_v}{(1 + \beta A_v)}$ where $\beta = \dfrac{1}{25}$.

Hence

$$G = \dfrac{\dfrac{2 \times 10^5}{(1 + j0.2f)}}{1 + \dfrac{2 \times 10^5}{25(1 + j0.2f)}} = \dfrac{50 \times 10^5}{25(1 + j0.2f) + 2 \times 10^5} = \dfrac{50 \times 10^5}{2 \times 10^5 + j5f}$$

$$= \dfrac{25}{1 + j2.5 \times 10^{-5} f}$$

The low-frequency gain is $1/\beta = 25$ ($= 28$ dB).

When $f = 15 \times 10^3$ Hz,	$G = 25/(1 + j0.375)$	$|G| = 23.4$ ($= 27.4$ dB).
When $f = 30 \times 10^3$ Hz,	$G = 25/(1 + j0.75)$	$|G| = 20$ ($= 26$ dB).
When $f = 60 \times 10^3$ Hz,	$G = 25/(1 + j1.5)$	$|G| = 13.9$ ($= 22.8$ dB).
When $f = 120 \times 10^3$ Hz,	$G = 25/(1 + j3)$	$|G| = 7.9$ ($= 18$ dB).
When $f = 240 \times 10^3$ Hz,	$G = 25/(1 + j6)$	$|G| = 4.1$ ($= 12.3$ dB).

These values are plotted in Figure 4.11. The 3 dB point is at a frequency of about 40 kHz.

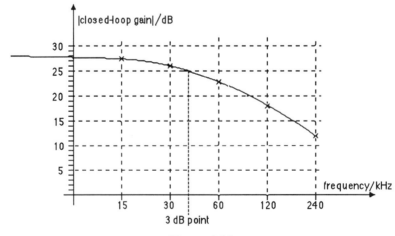

Figure 4.11

Solutions

The closed-loop gain-bandwidth product is therefore 25×40 kHz = 1 MHz.

The open-loop gain-bandwidth product is $2 \times 10^5 \times 5$ Hz which is also 1 MHz.

The gain-bandwidth product is unaffected by the feedback, as expected.

Question 4.5

(a)

Since negligible current flows in the input terminals of the amplifier (infinite input impedance), the voltage at the non-inverting input of the amplifier can be evaluated using the voltage divider rule as:

$$v_+ = V_2 \cdot \frac{R_2}{R_1 + R_2}$$

Since v_i is negligibly small, the voltages at the two input terminals of the amplifier must be equal, so

$$v_- = V_2 \cdot \frac{R_2}{R_1 + R_2}$$

At the non-inverting input of the amplifier, since the input current to the amplifier is negligible, the current through the input resistor R_1 must equal the current flowing in the feedback resistor R_2. So,

$$\frac{V_1 - v_-}{R_1} = \frac{v_- - V_o}{R_2}$$

$$\frac{V_1}{R_1} - v_-\left(\frac{1}{R_1} + \frac{1}{R_2}\right) = -\frac{V_o}{R_2}$$

$$\frac{V_1}{R_1} - v_-\left(\frac{R_1 + R_2}{R_1 R_2}\right) = -\frac{V_o}{R_2}$$

Substituting for v_-,

$$\frac{V_1}{R_1} - \frac{V_2}{R_1} = -\frac{V_o}{R_2}$$

and hence,

$$V_o = \frac{R_2}{R_1}(V_2 - V_1)$$

(b)

In the circuit of Figure 4.5, a convenient method of evaluating V_o is to use nodal analysis. The only node whose voltage is unknown is the junction of R_A, R_B and R_C, and writing the current law equation for this node gives:

Amplifiers and feedback

$$\frac{V_A - V_O}{R_A} + \frac{V_B - V_O}{R_B} = \frac{V_O}{R_C}$$

$$\frac{V_A}{R_A} + \frac{V_B}{R_B} = V_O\left[\frac{1}{R_A} + \frac{1}{R_B} + \frac{1}{R_C}\right] = V_O\frac{1}{R_E}$$

Hence, $$V_O = R_E\left(\frac{V_A}{R_A} + \frac{V_B}{R_B}\right)$$

(c)

In the circuit of Figure 4.6, since all resistors are of equal value, using the result of part (b) gives:

$$v_- = v_+ = \frac{1}{3}(V_2 + V_3)$$

and at the non-inverting input,

$$\frac{V_1 - v_-}{R} = \frac{v_- - V_o}{R}$$

$$V_o = 2v_- - V_1$$

$$V_o = \frac{2}{3}V_2 + \frac{2}{3}V_3 - V_1$$

Question 4.6

In the circuit of Figure 4.7,

$$G = -\frac{Z_F}{Z_S} = -\frac{20}{2(1 - j0.01\omega)} = -\frac{10}{(1 - j0.01\omega)}$$

At $\omega = 10$ rad s^{-1},

$$G = -\frac{10}{(1 - j0.1)} = -10\angle 5.7°$$

[This could equally be expressed as $+10\angle 185.7°$, but I prefer to retain the minus sign to reflect the inherent inversion in the feedback amplifier configuration.]

At $\omega = 100$ rad s^{-1}, $$G = -\frac{10}{(1 - j)} = -7.0\angle 45°$$

At $\omega = 1000$ rad s^{-1}, $$G = -\frac{10}{(1 - j10)} = -1\angle 84°$$

Question 4.7

(a)

The input circuit of the feedback amplifier of Figure 4.8 is shown in Figure 4.12, together with the Thévenin equivalent circuit. You are required to obtain values for V_T and Z_T.

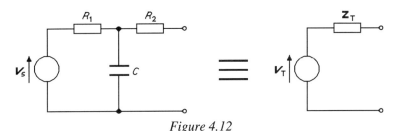

Figure 4.12

Clearly, since there is no voltage drop in R_2 when there is no load, the value of V_T is given by

$$V_T = V_S \times \frac{1/j\omega C}{R_1 + 1/j\omega C} = V_S \times \frac{1}{j\omega C R_1 + 1} = V_S \times \frac{1}{1+j\omega T_1}$$

Z_T will consist of R_2 in series with the parallel combination of R_1 and C.

$$Z_T = R_2 + \frac{R_1 \times 1/j\omega C}{R_1 + 1/j\omega C} = R_2 + \frac{R_1}{j\omega C R_1 + 1}$$

$$= \frac{j\omega C R_1 R_2 + R_2 + R_1}{j\omega C R_1 + 1}$$

$$= (R_1 + R_2) \times \frac{1 + j\omega C R_1 R_2/(R_1 + R_2)}{1 + j\omega C R_1}$$

$$= (R_1 + R_2) \times \frac{1 + j\omega T_2}{1 + j\omega T_1}$$

Replacing the input circuit of Figure 4.8 with the Thévenin equivalent circuit of Figure 4.12 results in the circuit of Figure 4.13.

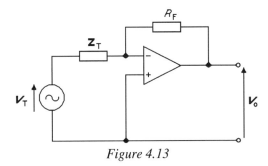

Figure 4.13

Amplifiers and feedback

Hence,

$$\frac{V_o}{V_T} = \frac{R_F}{Z_T} = \frac{R_F}{R_1 + R_2} \times \frac{1 + j\omega T_1}{1 + j\omega T_2}$$

$$\frac{V_o}{V_S} = \frac{V_o}{V_T} \times \frac{V_T}{V_S} = \frac{R_F}{R_1 + R_2} \times \frac{1 + j\omega T_1}{1 + j\omega T_2} \times \frac{1}{1 + j\omega T_1}$$

$$= \frac{R_F}{R_1 + R_2} \times \frac{1}{1 + j\omega T_2}$$

which is an expression of the required form.

(b)

For the given component values,

$$T_2 = \frac{CR_1 R_2}{R_1 + R_2} = \frac{0.5 \times 10^{-6} \times 25 \times 10^6}{10^4} \text{ s} = 1.25 \text{ ms}$$

$$G = \frac{V_o}{V_S} = \frac{200 \times 10^3}{10^4} \times \frac{1}{1 + j 1.25 \times 10^{-3} \omega} = \frac{20}{1 + j 1.25 \times 10^{-3} \omega}$$

The low-frequency gain = 20 (= 26 dB).

The break-point frequency = $1/(1.25 \times 10^{-3})$ rad s^{-1} = 800 rad s^{-1} = 127 Hz.

The straight-line approximation to the closed loop Bode gain plot is sketched in Figure 4.14.

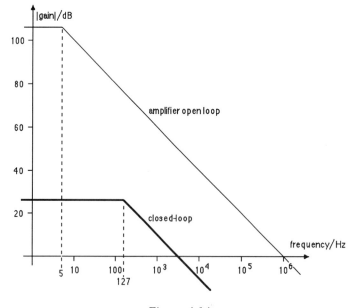

Figure 4.14

Solutions

(c)

Also plotted on Figure 4.14 is the amplifier's open-loop Bode gain plot.

For the virtual earth assumption to be valid, it is necessary that the loop gain $\mathbf{A}_v\beta$ shall be much greater than 1 over all frequencies of interest.

Expressing gains and the feedback fraction in dB,

or
$$20 \log_{10} |\mathbf{A}_v\beta| \gg 0 \text{ dB}$$
$$20 \log_{10} |\mathbf{A}_v| + 20 \log_{10} |\beta| \gg 0 \text{ dB}$$
and
$$20 \log_{10} |\mathbf{A}_v| - 20 \log_{10} |1/\beta| \gg 0 \text{ dB}$$

Since the closed-loop gain will be equal to $1/\beta$ provided $\mathbf{A}_v\beta$ is much greater than 1, for a first assessment of the validity of the virtual earth assumption we can look at the difference between the Bode gain plots of the amplifier gain ($20 \log_{10} |\mathbf{A}_v|$) and the closed-loop gain (assumed to be $20 \log_{10} |1/\beta|$. If this difference is significantly greater than 0 dB then $\mathbf{A}_v\beta$ is much greater than 1.

Looking at Figure 4.14, the difference between the two plots is at least 50 dB (i.e a value of $\mathbf{A}_v\beta$ of about 300) over all frequencies of interest, and hence the virtual earth assumption is valid over this frequency range.

Question 4.8

(a)

The equation for output voltage offset given in the textbook (page 187) is

$$V_o = (1 + R_F/R_1)V_{IO} + I_B[R_F - R_3(1 + R_F/R_1)] + I_{IO}[R_F + R_3(1 + R_F/R_1)]/2$$

For this example, $R_F = 120 \text{ k}\Omega$, $R_1 = 30 \text{ k}\Omega$ and R_3 is 2 kΩ.

Hence,

$$V_o = \left\{ \left(1 + \frac{120}{30}\right) \times 2 \times 10^{-3} + 150 \times 10^{-9} \times \left[120 - 2\left(1 + \frac{120}{30}\right)\right] \times 10^3 + 50 \times 10^{-9} \times \left[120 + 2\left(1 + \frac{120}{30}\right)\right] \times 10^3/2 \right\} \text{ V}$$

$$= \{10 \times 10^{-3} + 150 \times 10^{-6} \times 110 + 50 \times 10^{-6} \times 130/2\} \text{ V}$$

$$= \{10 + 16.5 + 3.25\} \text{ mV}$$

$$= 29.75 \text{ mV}$$

(b)

To minimise the contribution of I_B to the output offset voltage, the resistance R_3 should be equal to the parallel resistance of R_1 and R_F. In this case therefore, R_3 should be increased to 24 kΩ (30 kΩ in parallel with 120 kΩ) by adding a 22 kΩ resistor in series with the signal source. Since the input resistance of this non-inverting amplifier configuration is very high, the extra resistance will cause no significant change to the closed-loop gain of the amplifier.

Amplifiers and feedback

Question 4.9

(a)

The r.m.s. value of the total noise output voltage of an amplifier can be calculated as

$$V_{oN} = \sqrt{V_{oNV}^2 + V_{oNI}^2}$$

where V_{oNV} is the r.m.s. output noise voltage due to the equivalent input noise voltage generator v_{NA} and V_{oNI} is the r.m.s. output noise voltage due to the equivalent input noise current generator i_{NA}.

The expressions for V_{oNV} and V_{oNI} are:

$$V_{oNV} = (1 + R_F/R_1)V_{NA} \quad \text{and} \quad V_{oNI} = I_{NA}\left[R_F + R_3(1 + R_F/R_1)\right]$$

where V_{NA} is the r.m.s. value of v_{NA}, I_{NA} is the r.m.s. value of i_{NA}.

The value of V_{NA} is found from the graph of spectral density in Figure 4.31(a) of the textbook as follows.

From 10 Hz to 1 kHz, the average value of the graph is about 10^{-15} V² Hz⁻¹, so the contribution to V_{NA} of this portion of the frequency range is

$$\overline{v_{NA}^2}(1) = 10^{-15} \text{ V}^2 \text{ Hz}^{-1} \times 990 \text{ Hz} \approx 10^{-12} \text{ V}^2$$

From 1 kHz to 100 kHz, the value of the graph is constant at about 4×10^{-16} V² Hz⁻¹, so the contribution to V_{NA} of this portion of the frequency range is

$$\overline{v_{NA}^2}(2) = 4 \times 10^{-16} \text{ V}^2 \text{ Hz}^{-1} \times 99 \times 10^3 \text{ Hz} \approx 40 \times 10^{-12} \text{ V}^2$$

The total mean-square noise voltage over the frequency range 10 Hz to 100 kHz is the total area under the curve and is therefore the sum of these two contributions, so $v_{NA}^2 = 41 \times 10^{-12}$ V²

Hence the r.m.s. value is $V_{NA} = 6.4 \text{ µV}$

The value of I_{NA} is found from the graph of Figure 4.31(b) of the textbook in a similar manner.

From 10 Hz to 1 kHz the average value of the graph is about 7×10^{-24} A² Hz⁻¹ while from 1 kHz to 100 kHz the average value is about 5×10^{-25} A² Hz⁻¹. Hence

$$\overline{i_{NA}^2} = (7 \times 10^{-24} \text{ A}^2 \text{ Hz}^{-1} \times 990 \text{ Hz}) + (5 \times 10^{-25} \text{ A}^2 \text{ Hz}^{-1} \times 99 \text{ kHz})$$

$$\approx 7 \times 10^{-21} + 50 \times 10^{-21} \text{ A}^2$$

$$= 5.7 \times 10^{-20} \text{ A}^2$$

Hence $\quad I_{NA} = 2.4 \times 10^{-10}$ A $= 240$ pA

Solutions

The component values in the formulae for V_{oNV} and V_{oNI} are $R_1 = 10$ kΩ, $R_F = 1$ MΩ and $R_3 = 10$ kΩ in parallel with 1 M$\Omega \approx 10$ kΩ. Hence:

$$V_{oNV} = (1 + R_F/R_1)V_{NA} = 101 \times 6.4 \times 10^{-6} \text{ V} = 646 \text{ }\mu\text{V}$$

$$V_{oNI} = I_{NA}[R_F + R_3(1 + R_F/R_1)] = 240 \times 10^{-12} \times (10^6 + 10^4 \times 101) \text{ V}$$
$$= 482 \text{ }\mu\text{V}$$

Hence,
$$V_{oN} = \sqrt{V_{oNV}^2 + V_{oNI}^2} = \sqrt{482^2 + 646^2} = 806 \text{ }\mu\text{V}$$

(b)

As established in Question 4.5, the amplifier configuration of Figure 4.10 produces an output voltage which is 100 times the difference between V_1 and V_2, i.e 100 times the voltage v.

For a signal-to-noise ratio of 1000 or more at the amplifier output, the signal-related amplifier output voltage must be at least 806 mV. Since the amplifier gain is 100, the differential input voltage v must be at least 8.06 mV.

Question 4.10

(a)

The closed-loop gain of the amplifer of Figure 4.9 is 5, so the amplitude of the output voltage is 10 V.

At a signal frequency ω, $\qquad v_o = V_o \sin \omega t$

and $\qquad dv_o/dt = \omega V_o \cos \omega t$

Hence the maximum value of ω is given by

$$\omega_{max} = (dv_o/dt)_{max} / V_o = 10^7 \text{ V s}^{-1} / 10 \text{ V} = 10^6 \text{ rad s}^{-1}$$

Hence $\qquad f_{max} = \omega_{max}/2\pi \approx 160$ kHz

(b)

To amplify 1 MHz without distortion, the maximum output signal amplitude will be given by

$$(V_o)_{max} = 10^7/\omega = 10^7/(6.28 \times 10^6) \text{ V} = 1.6 \text{ V}$$

and hence the maximum input signal amplitude will be

$$(V_i)_{max} = (V_o)_{max}/\text{gain} = 1.6/5 \text{ V} = 0.32 \text{ V}$$

5 COMBINATIONAL LOGIC CIRCUITS

QUESTIONS

5.1 A combinational logic circuit is required which produces an output D from four input signals $A, B, C1$ and $C2$ according to the following rules.

    (i)   If $C1$ and $C2$ are both 1, the output D must be 0.

    (ii)   If $C1$ and $C2$ are both 0, the output D must be 1.

    (iii)   If $C1 = 1$ and $C2 = 0$, then output D must be equal to input A.

    (iv)   If $C1 = 0$ and $C2 = 1$, then output D must be equal to input B.

(a) Construct the truth table for D.

(b) Construct the Karnaugh map for D.

(c) By combining 1s in the largest possible groups on the Karnaugh map, obtain an expression for D in the simplest sum-of-products form.

(d) Draw the gate implementation of the logic circuit to generate the output D using AND gates, OR gates and inverters.

5.2 Two 2-bit binary numbers AB and CD, when multiplied together, produce a 4-bit binary product $EFGH$, where each binary digit has its usual positional weighting.

(a) Complete the truth table of Table 5.1.

A	B	C	D	E	F	G	H
0	0	0	0				
0	0	0	1				
0	0	1	0				
0	0	1	1	0	0	0	0
0	1	0	0				
0	1	0	1				
0	1	1	0				
0	1	1	1	0	0	1	1
1	0	0	0				
1	0	0	1				
1	0	1	0				
1	0	1	1	0	1	1	0
1	1	0	0				
1	1	0	1				
1	1	1	0				
1	1	1	1				

Table 5.1

Combinational logic circuits

(b) Construct the Karnaugh map for each of the four output variables.

(c) From the Karnaugh maps, obtain an expression for each of the outputs in the simplest possible sum-of-products form, and draw the circuit to generate each output using only NAND gates and inverters.

(d) From the same Karnaugh maps, by grouping together 0s, obtain an expression for the complement of each output, and hence draw the circuit to generate each output using only AND and NOR gates. (You may assume that the complement of each input variable is available, as well as the variable itself.)

5.3 A logic circuit is to be designed which has, as inputs, two 2-bit binary numbers, AB and CD. The circuit is required to generate three output signals, G, L and E. G is to be 1 only when the number AB is greater than the number CD, L is to be 1 only when the number AB is less than the number CD, while E is to be 1 only when the two input numbers are equal.

(a) Construct the truth table with AB and CD as inputs and G, L and E as outputs.

(b) Construct the Karnaugh maps for each of the output variables.

(c) Use the Karnaugh maps to obtain an expression for each output variable in the simplest possible sum-of-products form.

(d) By re-arranging the expression for E using the commutative and associative laws, show how this variable could be generated from the input variables using two XNOR-gates and one AND-gate.

5.4 The Gray code is a binary code often used with transducers which generate a digital signal proportional to rotary or linear movement. A four-bit Gray code has the form shown in Table 5.2

A	B	C	D	denary value
0	0	0	0	0
0	0	0	1	1
0	0	1	1	2
0	0	1	0	3
0	1	1	0	4
0	1	1	1	5
0	1	0	1	6
0	1	0	0	7
1	1	0	0	8
1	1	0	1	9
1	1	1	1	10
1	1	1	0	11
1	0	1	0	12
1	0	1	1	13
1	0	0	1	14
1	0	0	0	15

Table 5.2

(a) You are required to design a Gray code to binary code converter. As a first step, complete the truth table of Table 5.3.

input Gray code				output binary code			
A	B	C	D	a	b	c	d
0	0	0	0				
0	0	0	1				
0	0	1	0				
0	0	1	1				
0	1	0	0				
0	1	0	1				
0	1	1	0				
0	1	1	1				
1	0	0	0				
1	0	0	1				
1	0	1	0				
1	0	1	1				
1	1	0	0				
1	1	0	1				
1	1	1	0				
1	1	1	1				

Table 5.3

(b) Construct the Karnaugh map for each of the output variables and, from these maps, write an expression for each variable in the simplest possible sum-of-products form.

(c) By selecting groups of zeros, write an expression for the complement of each output variable in the simplest possible sum-of-products form.

5.5 (a) Draw the gate implementation of the logic functions for b and c derived in Question 5.4 part (b), using AND-gates, OR-gates and inverters.

(b) Draw the NAND-gate implementation of the same logic functions.

5.6 (a) Make the required additions to the simplified logic diagram of the PAL 14L8 in Figure 5.1 to show how the code converter of Questions 5.4 and 5.5 could be implemented using just one PAL 14L8.

[*Hint*: you may find it more convenient to think about using the complement functions derived in Question 5.4 part (c) as part of your solution.]

(b) Using the relationships obtained in Question 5.4 parts (b) and (c), show how the logic functions for b, c and d can be manipulated using the commutative, associative and distributive laws to establish that all three variables can be generated from the input variables A, B, C and D using only three 2-input XOR-gates, and draw the required gate implementation.

Combinational logic circuits

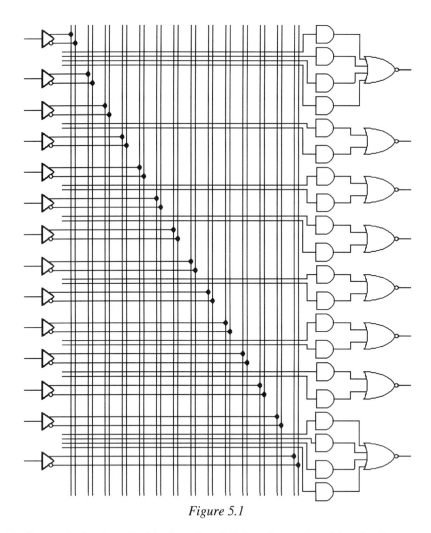

Figure 5.1

5.7 A 4-bit Gray code (as described in Question 5.4) is to be used to drive the 7-segment indicator of Figure 5.2(a). The display to be produced by each of the 16 possible input codes is shown in Figure 5.2(b). Asume that a logic 1 input is required to illuminate a segment.

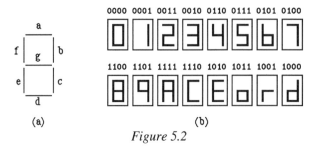

Figure 5.2

(a) Construct the truth table showing all possible states of the input code and the 7 output signals to be generated by the logic circuit required between the Gray code input and the display device.

(b) From the truth table, construct the Karnaugh map for each of the output variables.

5.8 For the design exercise of Question 5.7,

(a) Obtain an expression for each of the output variables in the simplest possible sum-of-products form.

(b) Draw the NAND-gate implementation of the logic circuit for each of the output variables.

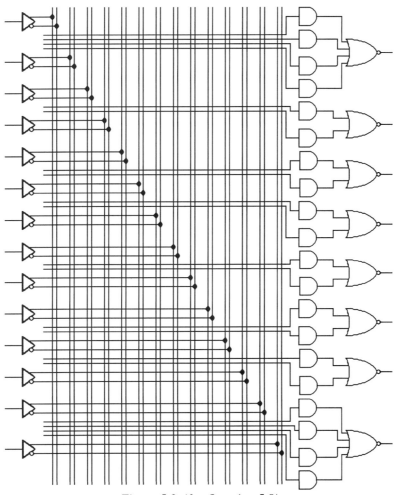

Figure 5.3 (for Question 5.9)

Combinational logic circuits

5.9 Could the complete logic circuit of the solution to Question 5.8 be implemented using one PAL 14L8? If so, show, on Figure 5.3, the way in which the device would be used. If not, work out how many PAL 14L8s would be required and indicate on Figure 5.3 how you would obtain maximum possible use of one such device.

5.10 The design of Questions 5.7, 5.8 and 5.9 is to be modified on the assumption that only the Gray codes corresponding to the denary numbers 0 to 9 will ever occur as inputs, and that therefore the display generated by the unused codes is immaterial.

Repeat the design process of Question 5.7 parts (a) and (b) and Question 5.9 for the new situation. Another blank diagram for the PAL 14L8 is provided as Figure 5.4 for your answer.

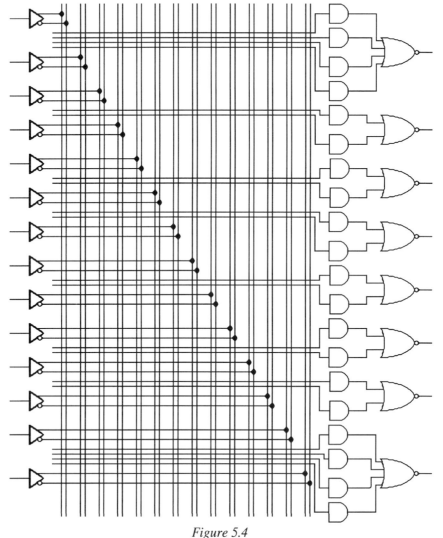

Figure 5.4

SOLUTIONS

Question 5.1

(a) The required truth table is shown in Table 5.4.

A	B	C1	C2	D
0	0	0	0	1
0	0	0	1	0
0	0	1	0	0
0	0	1	1	0
0	1	0	0	1
0	1	0	1	1
0	1	1	0	0
0	1	1	1	0
1	0	0	0	1
1	0	0	1	0
1	0	1	0	1
1	0	1	1	0
1	1	0	0	1
1	1	0	1	1
1	1	1	0	1
1	1	1	1	0

Table 5.4

(b) The Karnaugh map is shown in Figure 5.5, with the chosen groups of 1s.

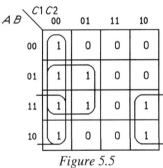

Figure 5.5

(c) From the Karnaugh map, $D = A.\overline{C2} + B.\overline{C1} + \overline{C1}.\overline{C2}$

(d) The logic circuit is shown in Figure 5.6

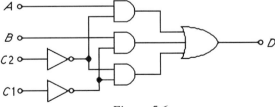

Figure 5.6

Combinational logic circuits

Question 5.2

(a) The complete truth table is shown in Table 5.5

A	B	C	D	E	F	G	H
0	0	0	0	0	0	0	0
0	0	0	1	0	0	0	0
0	0	1	0	0	0	0	0
0	0	1	1	0	0	0	0
0	1	0	0	0	0	0	0
0	1	0	1	0	0	0	1
0	1	1	0	0	0	1	0
0	1	1	1	0	0	1	1
1	0	0	0	0	0	0	0
1	0	0	1	0	0	1	0
1	0	1	0	0	1	0	0
1	0	1	1	0	1	1	0
1	1	0	0	0	0	0	0
1	1	0	1	0	0	1	1
1	1	1	0	0	1	1	0
1	1	1	1	1	0	0	1

Table 5.5

(b) The Karnaugh maps are shown in Figure 5.7

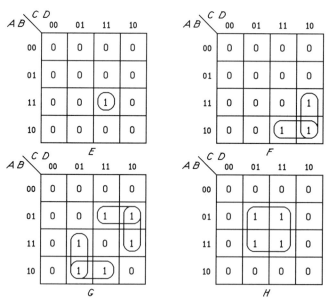

Figure 5.7

(c) The logic functions are:

$$E = A.B.C.D$$
$$F = A.\overline{B}.C + A.C.\overline{D}$$
$$G = \overline{A}.B.C + A.\overline{B}.D + A.\overline{C}.D + B.C.\overline{D}$$
$$H = B.D$$

The logic circuit to generate the output signals is shown in Figure 5.8.

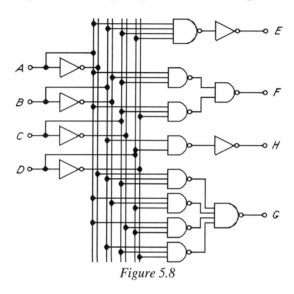

Figure 5.8

(d)

The Karnaugh maps with grouped 0s are shown in Figure 5.9

The resulting expressions for the complements of the output variables are:

$$\overline{E} = \overline{A} + \overline{B} + \overline{C} + \overline{D}$$
$$\overline{F} = \overline{A} + \overline{C} + B.D$$
$$\overline{G} = \overline{A}.\overline{B} + \overline{A}.\overline{C} + \overline{B}.\overline{D} + \overline{C}.\overline{D} + A.B.C.D$$
$$\overline{H} = \overline{B} + \overline{D}$$

which give the following expressions for the output variables:

$$E = \overline{\overline{A} + \overline{B} + \overline{C} + \overline{D}}$$
$$F = \overline{\overline{A} + \overline{C} + B.D}$$
$$G = \overline{\overline{A}.\overline{B} + \overline{A}.\overline{C} + \overline{B}.\overline{D} + \overline{C}.\overline{D} + A.B.C.D}$$
$$H = \overline{\overline{B} + \overline{D}}$$

Combinational logic circuits

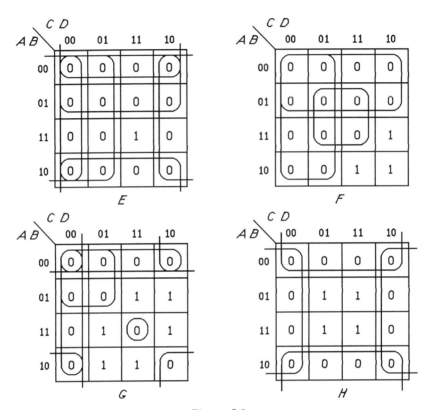

Figure 5.9

The circuit to implement these functions using only AND and NOR gates is shown in Figure 5.10

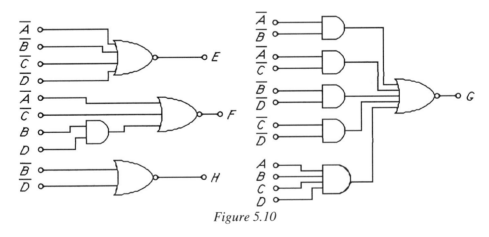

Figure 5.10

Question 5.3

(a)

The required truth table is shown in Table 5.6.

A	B	C	D	G	L	E
0	0	0	0	0	0	1
0	0	0	1	0	1	0
0	0	1	0	0	1	0
0	0	1	1	0	1	0
0	1	0	0	1	0	0
0	1	0	1	0	0	1
0	1	1	0	0	1	0
0	1	1	1	0	1	0
1	0	0	0	1	0	0
1	0	0	1	1	0	0
1	0	1	0	0	0	1
1	0	1	1	0	1	0
1	1	0	0	1	0	0
1	1	0	1	1	0	0
1	1	1	0	1	0	0
1	1	1	1	0	0	1

Table 5.6

(b)

The Karnaugh map for each of the output variables is shown in Figure 5.11.

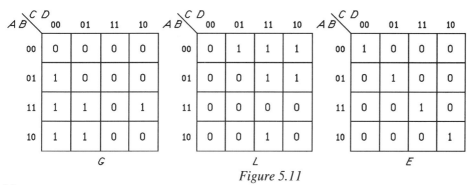

Figure 5.11

(c)

The simplest possible sum-of-product expressions for the outputs are:

$$G = A.\overline{C} + A.B.\overline{D} + B.\overline{C}.\overline{D}$$

$$L = \overline{A}.C + \overline{A}.\overline{B}.D + \overline{B}.C.D$$

$$E = \overline{A}.\overline{B}.\overline{C}.\overline{D} + \overline{A}.B.\overline{C}.D + A.B.C.D + A.\overline{B}.C.\overline{D}$$

Combinational logic circuits

(d)

The expression for E can be re-arranged as follows:

$$E = \overline{A}.\overline{B}.\overline{C}.\overline{D} + \overline{A}.B.\overline{C}.D + A.B.C.D + A.\overline{B}.C.\overline{D}$$
$$= \overline{A}.\overline{C}(\overline{B}.\overline{D} + B.D) + A.C(\overline{B}.\overline{D} + B.D)$$
$$= (\overline{A}.\overline{C} + A.C).(\overline{B}.\overline{D} + B.D)$$
$$= \overline{(A \oplus C)}.\overline{(B \oplus D)}$$

The gate implementation of this expression is shown in Figure 5.12

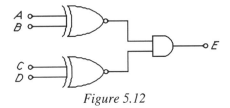

Figure 5.12

Question 5.4

(a)

The completed truth table is shown in Table 5.7.

input Gray code				output binary code			
A	B	C	D	a	b	c	d
0	0	0	0	0	0	0	0
0	0	0	1	0	0	0	1
0	0	1	0	0	0	1	1
0	0	1	1	0	0	1	0
0	1	0	0	0	1	1	1
0	1	0	1	0	1	1	0
0	1	1	0	0	1	0	0
0	1	1	1	0	1	0	1
1	0	0	0	1	1	1	1
1	0	0	1	1	1	1	0
1	0	1	0	1	1	0	0
1	0	1	1	1	1	0	1
1	1	0	0	1	0	0	0
1	1	0	1	1	0	0	1
1	1	1	0	1	0	1	1
1	1	1	1	1	0	1	0

Table 5.7

Solutions

(b)

It is clear from the truth table that $a = A$.

The Karnaugh maps for the other three output variables are shown in Figure 5.13.

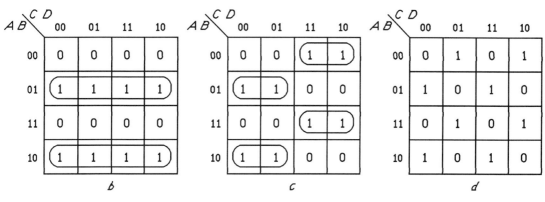

Figure 5.13

From these maps, using the groups shown,

$b = \overline{A}.B + A.\overline{B}$

$c = \overline{A}.\overline{B}.C + \overline{A}.B.\overline{C} + A.B.C + A.\overline{B}.\overline{C}$

$d = \overline{A}.\overline{B}.\overline{C}.D + \overline{A}.\overline{B}.C.\overline{D} + \overline{A}.B.\overline{C}.\overline{D} + \overline{A}.B.C.D + A.B.\overline{C}.D + A.B.C.\overline{D} + A.\overline{B}.\overline{C}.\overline{D} + A.\overline{B}.C.D$

(c)

Using groups of 0s instead of groups of 1s we get,

$\overline{b} = \overline{A}.\overline{B} + A.B$

$\overline{c} = \overline{A}.\overline{B}.\overline{C} + \overline{A}.B.C + A.B.\overline{C} + A.\overline{B}.C$

$\overline{d} = \overline{A}.\overline{B}.\overline{C}.\overline{D} + \overline{A}.\overline{B}.C.D + \overline{A}.B.\overline{C}.D + \overline{A}.B.C.\overline{D} + A.B.\overline{C}.\overline{D} + A.B.C.D + A.\overline{B}.\overline{C}.D + A.\overline{B}.C.\overline{D}$

Question 5.5

(a)

The AND, OR and INVERTER gate implementation of the logic functions for b and c found in Question 5.4 part (b) is shown in Figure 5.14.

Combinational logic circuits

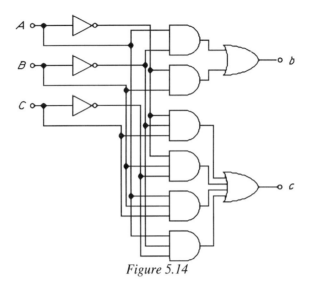

Figure 5.14

(b)

The NAND-gate logic implementation of the variables b and c is shown in Figure 5.15

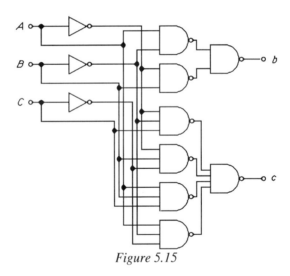

Figure 5.15

This can be deduced by using deMorgan's theorem as follows:

If $Z = V.W + X.Y$ then $\overline{Z} = \overline{V.W}.\overline{X.Y}$ and $Z = \overline{\overline{Z}} = \overline{\overline{V.W}.\overline{X.Y}}$ which can be implemented using only NAND-gates.

Question 5.6

(a)

Figure 5.16 shows one possible implementation of the code converter using a PAL 14L8. There are other valid configurations. (I have deliberately omitted the connection which shows that $a = A$ because it does not involve the PAL device.)

This solution is arrived at as follows.

Using the relationship developed in Question 5.4 part (c),

$$\bar{b} = \bar{A}.\bar{B} + A.B \quad \text{so} \quad b = \overline{\bar{A}.\bar{B} + A.B}$$

which represents the output of the uppermost 2-input NOR-gate.

Similarly, since $\quad \bar{c} = \bar{A}.\bar{B}.\bar{C} + \bar{A}.B.C + A.B.\bar{C} + A.\bar{B}.C,$

$$c = \overline{\bar{A}.\bar{B}.\bar{C} + \bar{A}.B.C + A.B.\bar{C} + A.\bar{B}.C}$$

which is the output of the upper 4-input NOR-gate.

The function for d is less easily implemented, since there is no provision of an 8-input NOR-gate in the device, and hence the product terms must be grouped together before being input to the final NOR-gate.

I have chosen to combine the product terms into four pairs as shown below.

$$\bar{d} = \left(\bar{A}.\bar{B}.\bar{C}.\bar{D} + \bar{A}.\bar{B}.C.D\right) + \left(\bar{A}.B.\bar{C}.D + \bar{A}.B.C.\bar{D}\right)$$
$$+ \left(A.B.\bar{C}.\bar{D} + A.B.C.D\right) + \left(A.\bar{B}.\bar{C}.D + A.\bar{B}.C.\bar{D}\right)$$
$$= W + X + Y + Z$$

So, $d = \overline{W + X + Y + Z}$ will be generated by the lower 4-input NOR-gate. The inputs to this NOR-gate need to be W, X, Y and Z, but the 2-input NOR-gates generate $\bar{W}$, $\bar{X}$, $\bar{Y}$ and $\bar{Z}$, so these terms must be fed through inverters before being fed, via single-input AND-gates to the final NOR-gate, as shown in the figure.

Combinational logic circuits

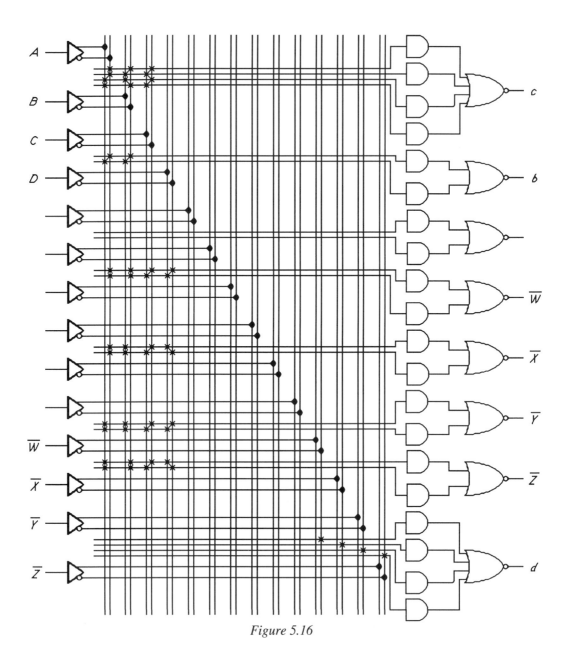

Figure 5.16

(b)

The expression for *b* derived in Question 5.4 part (b) is:

$$b = \bar{A}.B + A.\bar{B} \text{ which can also be written } b = A \oplus B$$

The expression for *c* is:

$$c = \bar{A}.\bar{B}.C + \bar{A}.B.\bar{C} + A.B.C + A.\bar{B}.\bar{C}$$

which can be re-arranged as:

$$c = (\bar{A}.\bar{B} + A.B).C + (\bar{A}.B + A.\bar{B}).\bar{C} = \bar{b}.C + b.\bar{C} = b \oplus C$$

The expression:

$$d = \bar{A}.\bar{B}.\bar{C}.D + \bar{A}.\bar{B}.C.\bar{D} + \bar{A}.B.\bar{C}.\bar{D} + \bar{A}.B.C.D + A.B.\bar{C}.D + A.B.C.\bar{D} + A.\bar{B}.\bar{C}.\bar{D} + A.\bar{B}.C.D$$

can also be re-arranged in a similar manner to give:

$$d = (\bar{A}.\bar{B}.\bar{C} + \bar{A}.B.C + A.B.\bar{C} + A.\bar{B}.C).D + (\bar{A}.\bar{B}.C + \bar{A}.B.\bar{C} + A.B.C + A.\bar{B}.\bar{C}).\bar{D}$$

$$= \bar{c}.D + c.\bar{D} = c \oplus D$$

Hence the complete decoder can be implemented by the circuit of Figure 5.17.

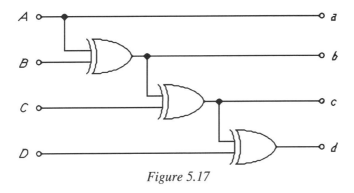

Figure 5.17

Question 5.7

(a)

The complete truth table is shown in Table 5.8. Note that the input codes have been written in the order of the denary numbers which they represent, rather than in pure binary order, which previous truth tables have used. This means that a little more care is required in constructing the resulting Karnaugh maps. For your convenience in checking your solution, Table 5.9 contains the same information, but with the input variable states listed in pure binary order.

Combinational logic circuits

denary	Gray code				indicator segments						
	A	B	C	D	a	b	c	d	e	f	g
0	0	0	0	0	1	1	1	1	1	1	0
1	0	0	0	1	0	1	1	0	0	0	0
2	0	0	1	1	1	1	0	1	1	0	1
3	0	0	1	0	1	1	1	1	0	0	1
4	0	1	1	0	0	1	1	0	0	1	1
5	0	1	1	1	1	0	1	1	0	1	1
6	0	1	0	1	0	0	1	1	1	1	1
7	0	1	0	0	1	1	1	0	0	0	0
8	1	1	0	0	1	1	1	1	1	1	1
9	1	1	0	1	1	1	1	0	0	1	1
10	1	1	1	1	1	1	1	0	1	1	1
11	1	1	1	0	1	0	0	1	1	1	0
12	1	0	1	0	1	0	0	1	1	1	1
13	1	0	1	1	0	0	1	1	1	0	1
14	1	0	0	1	0	0	0	0	1	0	1
15	1	0	0	0	0	1	1	1	1	0	1

Table 5.8

denary	Gray code				indicator segments						
	A	B	C	D	a	b	c	d	e	f	g
0	0	0	0	0	1	1	1	1	1	1	0
1	0	0	0	1	0	1	1	0	0	0	0
3	0	0	1	0	1	1	1	1	0	0	1
2	0	0	1	1	1	1	0	1	1	0	1
7	0	1	0	0	1	1	1	0	0	0	0
6	0	1	0	1	0	0	1	1	1	1	1
4	0	1	1	0	0	1	1	0	0	1	1
5	0	1	1	1	1	0	1	1	0	1	1
15	1	0	0	0	0	1	1	1	1	0	1
14	1	0	0	1	0	0	0	0	1	0	1
12	1	0	1	0	1	0	0	1	1	1	1
13	1	0	1	1	0	0	1	1	1	0	1
8	1	1	0	0	1	1	1	1	1	1	1
9	1	1	0	1	1	1	1	0	0	1	1
11	1	1	1	0	1	0	0	1	1	1	0
10	1	1	1	1	1	1	1	0	1	1	1

Table 5.9

Solutions

(b)

The Karnaugh maps are shown in Figure 5.18.

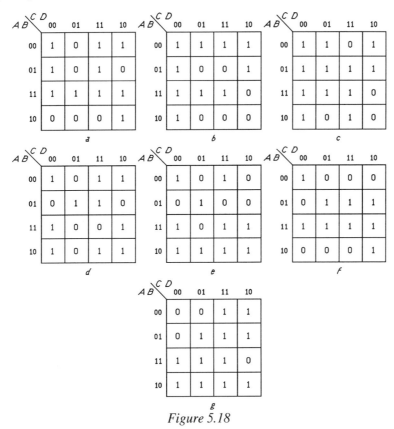

Figure 5.18

Question 5.8

(a)

Possible sum-of-products expressions for the output variables (there are alternatives) are:

$$a = A.B + \overline{A}.\overline{C}.\overline{D} + \overline{A}.C.D + \overline{B}.C.\overline{D}$$
$$b = \overline{A}.\overline{B} + \overline{C}.\overline{D} + \overline{A}.D + A.B.D$$
$$c = \overline{A}.B + \overline{A}.\overline{C} + \overline{A}.D + B.D + \overline{C}.\overline{D} + A.C.D$$
$$d = A.\overline{D} + \overline{B}.C + \overline{B}.\overline{D} + \overline{A}.B.D$$
$$e = A.\overline{B} + A.C + A.\overline{D} + \overline{B}.\overline{C}.\overline{D} + \overline{B}.C.D + \overline{A}.B.\overline{C}.D$$
$$f = A.B + B.C + B.D + A.C.\overline{D} + \overline{A}.\overline{B}.\overline{C}.\overline{D}$$
$$g = A.\overline{C} + \overline{A}.C + \overline{B}.C + B.D$$

Combinational logic circuits

(b)

The NAND-gate implementations of the logic circuits for the output variables are shown in Figure 5.19. The full gate requirement for each variable has been shown, but in practice, gates producing the same ouput function would probably not be duplicated. For example, there would only be 4 inverters, one for each input signal (provided the gate fan-out was not exceeded); the function $\overline{A.B}$ might be generated only once, even though it is required for both *a* and *f*, etc.

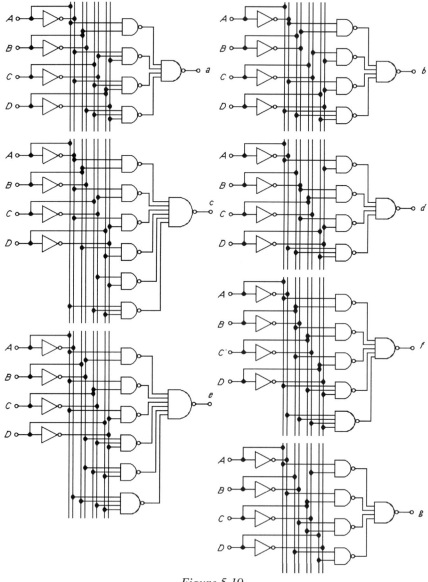

Figure 5.19

Question 5.9

As in Question 5.6 part (a), the best approach to the use of the PAL 14L8 device is, because of the NOR gate outputs, to start with the logic functions for the complement of each output variable. This means extracting the 0s from the Karnaugh maps.

From the maps of Figure 5.18, the required expressions are:

$$\bar{a} = A.\bar{B}.\bar{C} + A.\bar{B}.D + \bar{A}.\bar{C}.D + \bar{A}.B.C.\bar{D}$$
$$\bar{b} = \bar{A}.B.D + A.\bar{B}.D + A.C.\bar{D}$$
$$\bar{c} = A.C.\bar{D} + A.\bar{B}.\bar{C}.D + \bar{A}.\bar{B}.C.D$$
$$\bar{d} = A.B.D + \bar{A}.B.\bar{D} + \bar{B}.\bar{C}.D$$
$$\bar{e} = \bar{A}.B.C + \bar{A}.B.\bar{D} + \bar{A}.C.\bar{D} + A.B.\bar{C}.D + \bar{A}.\bar{B}.\bar{C}.D$$
$$\bar{f} = \bar{B}.D + A.\bar{B}.\bar{C} + \bar{A}.\bar{B}.C + \bar{A}.B.\bar{C}.\bar{D}$$
$$\bar{g} = \bar{A}.\bar{B}.\bar{C} + \bar{A}.\bar{C}.\bar{D} + A.B.C.\bar{D}$$

Clearly, there is a requirement for two 4-input NOR gates (to generate a and f), five 3-input NOR gates and one 5-input NOR gate. However, as in Question 5.3, functions can be split into pairs of product terms to enable 2-input NOR gates to be used. For example, the expression for $\bar{b}$ can be re-arranged as:

$$\bar{b} = (\bar{A}.B.D + A.\bar{B}.D) + A.C.\bar{D} \quad \text{or} \quad b = \overline{(\bar{A}.B.D + A.\bar{B}.D) + A.C.\bar{D}}$$

which can be implemented using two 2-input NOR gates and one inverter.

The expression for $\bar{e}$ can be re-arranged as:

$$\bar{e} = ((\bar{A}.B.C + \bar{A}.B.\bar{D}) + \bar{A}.C.\bar{D}) + (A.B.\bar{C}.D + \bar{A}.\bar{B}.\bar{C}.D)$$
$$e = \overline{((\bar{A}.B.C + \bar{A}.B.\bar{D}) + \bar{A}.C.\bar{D}) + (A.B.\bar{C}.D + \bar{A}.\bar{B}.\bar{C}.D)}$$

which can be implemented using four 2-input NOR gates and three inverters.

So, to implement the whole logic circuit would require two 4-input NOR gates, twelve 2-input NOR gates, 11 inverters (4 for the input variables and 7 for the NOR-OR conversions) plus the requisite number of AND gates (32 in this case).

One PAL 14L8 is therefore not able to implement the whole circuit, and another one would be required.

One PAL 14L8 could, however, implement the circuit to generate five of the required output variables, as shown in Figure 5.20.

Combinational logic circuits

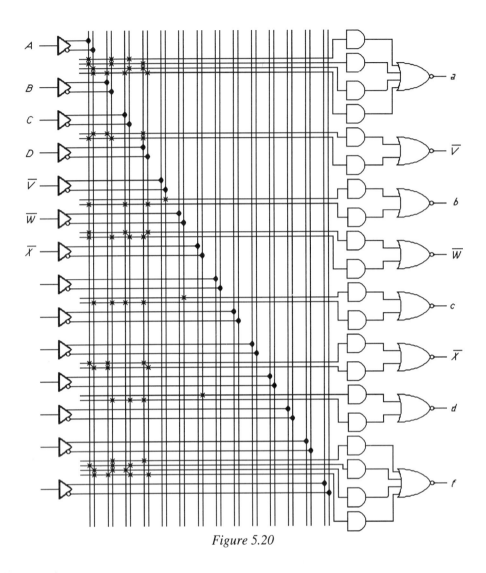

Figure 5.20

Question 5.10

The truth table for the new problem will be the same as that of Table 5.8 except that the last 6 lines of the table will contain all "don't care" conditions.

The Karnaugh maps for the output variables will be similar to those of Figure 5.18, but will have Xs (representing the don't care conditions) in six squares, as shown in Figure 5.21.

Selecting groups of 0s to obtain the complement of each of the output variables, as shown in Figure 5.21 gives the following expressions.

Solutions

$$\bar{a} = \bar{A}.\bar{C}.D + B.C.\bar{D}$$
$$\bar{b} = \bar{A}.B.D$$
$$\bar{c} = \bar{B}.C.D$$
$$\bar{d} = A.D + \bar{A}.B.\bar{D} + \bar{B}.\bar{C}.D$$
$$\bar{e} = A.D + B.C + C.\bar{D} + \bar{A}.B.\bar{D} + \bar{B}.\bar{C}.D$$
$$\bar{f} = \bar{B}.C + \bar{B}.D + \bar{A}.B.\bar{C}.\bar{D}$$
$$\bar{g} = \bar{B}.\bar{C} + \bar{A}.\bar{C}.\bar{D}$$

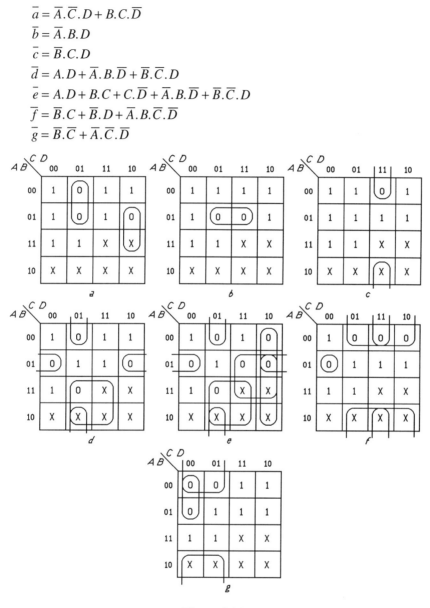

Figure 5.21

Can these 7 expressions be implemented using only one PAL 14L8? Variables a and g clearly each need a 2-input NOR gate, e will require a 4-input NOR plus a 2-input NOR, d can be implemented using two 2-input NORs while f can use the other 4-input NOR. Since b and c can only be generated by each utilising a 2-input NOR (even though they contain no NOR function) because the only chip outputs are from NOR gates, it seems as if the circuit needs one more 2-input NOR gate than the

Combinational logic circuits

PAL 14L8 contains. However, if you look carefully at the expressions, you will see that $\bar{e}$ contains the three terms $(A.D + \bar{A}.B.\bar{D} + \bar{B}.\bar{C}.D)$ which is the complement of d. So d and f can be generated using the two 4-input NOR gates, while e can be generated using two 2-input NOR gates plus two inverters as follows:

$$\bar{e} = A.D + B.C + C.\bar{D} + \bar{A}.B.\bar{D} + \bar{B}.\bar{C}.D$$

$$= \bar{d} + B.C + C.\bar{D}$$

$$= \bar{d} + \overline{\overline{B.C + C.\bar{D}}}$$

So, $e = \overline{\bar{d} + \overline{B.C + C.\bar{D}}}$

The complete circuit can therefore be implemented using one PAL 14L8, as shown in Figure 5.22.

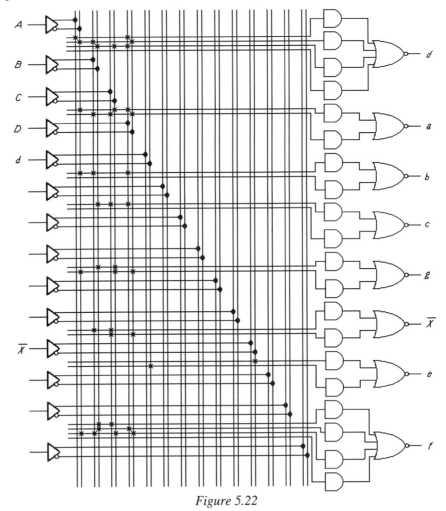

Figure 5.22

6 SEQUENTIAL LOGIC CIRCUITS

QUESTIONS

6.1 **(a)** Figure 6.1 shows the timing diagram of signals applied to the S and R inputs of the controlled bistable circuit of Figure 6.1 of the textbook (page 261). Assuming Q is initially 1, sketch on Figure 6.1 below the waveforms of the Q and $\overline{Q}$ outputs.

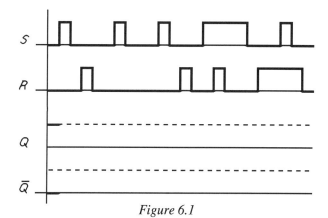

Figure 6.1

(b) Show that the circuit of Figure 6.2 is also a controlled bistable circuit and identify the functions of the A and B inputs. Complete the timing diagram of Figure 6.3 to show the variation in Q and $\overline{Q}$.

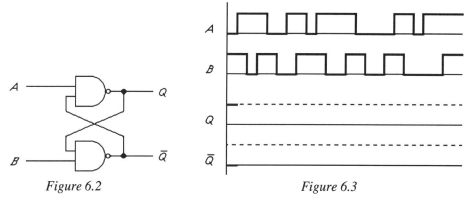

Figure 6.2 *Figure 6.3*

6.2 **(a)** Using the results of Question 6.1(b), deduce the circuit of a clocked RS latch using only NAND gates.

(b) Complete the timing diagram of Figure 6.4 to show the changes occurring in Q and $\overline{Q}$ of this new clocked RS latch.

Sequential logic circuits

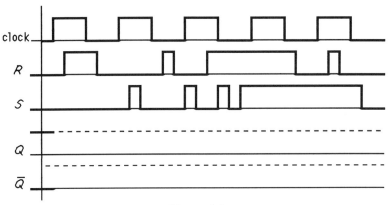

Figure 6.4

6.3 For the circuit of Figure 6.5, complete the timing diagram of Figure 6.6. Describe the function of the circuit.

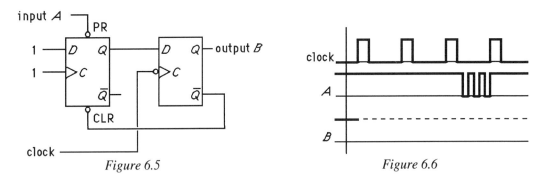

Figure 6.5   Figure 6.6

6.4 The circuit of Figure 6.7 has all flip-flops initially set to 1. It is then fed with clock pulses until the flip-flop ouputs are once more all 1s. Complete Table 6.1 to show all possible states of the output word

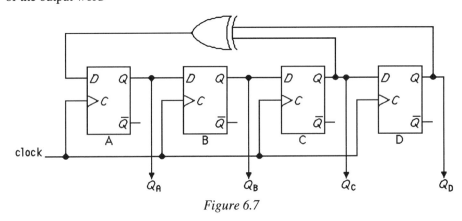

Figure 6.7

Questions

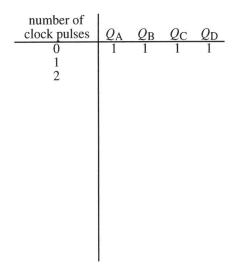

Table 6.1

6.5 **(a)** Figure 6.29 of the textbook shows a 3-bit natural binary ripple up-counter. From the description of that circuit given in the textbook, deduce and explain the operation of the circuit of Figure 6.8.

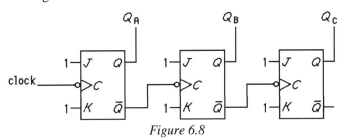

Figure 6.8

(b) Figure 6.9 shows a proposed 3-bit natural binary ripple counter. What is the intended function of the input X? By completing the timing diagram of Figure 6.10, deduce whether the circuit behaves in the way in which the designer intended.

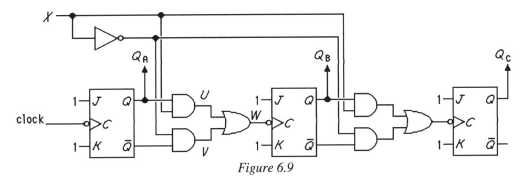

Figure 6.9

Sequential logic circuits

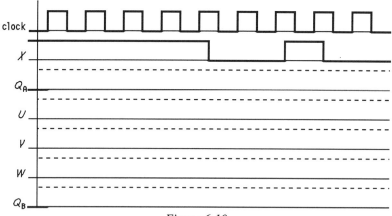

Figure 6.10

6.6 (a) Complete Table 6.2 for the circuit of Figure 6.11 and hence complete the timing diagram of Figure 6.12 relating the waveforms at $Q1$ and $Q2$ to the symmetrical square-wave clock signal.

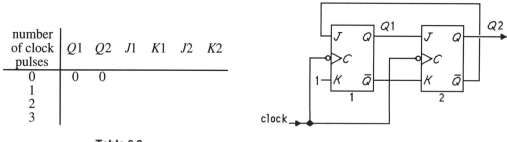

number of clock pulses	Q1	Q2	J1	K1	J2	K2
0	0	0				
1						
2						
3						

Table 6.2

Figure 6.11

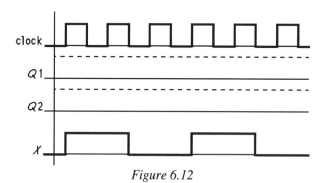

Figure 6.12

(b) By examining the timing diagram, deduce (i) the NOR-gate logic circuit and (ii) the NAND-gate logic circuit which, using the flip-flop outputs and the clock as inputs, can produce a symmetrical output square wave at one third of the clock frequency as shown at the bottom of Figure 6.12 (signal X).

Questions

6.7 **(a)** By using a table and timing diagram similar to those used in Question 6.6(a), deduce the way in which the ouputs Q_A, Q_B and Q_C of the circuit of Figure 6.13 are related to the input clock waveform. (Assume that the initial state of the circuit has all Q outputs equal to 0.)

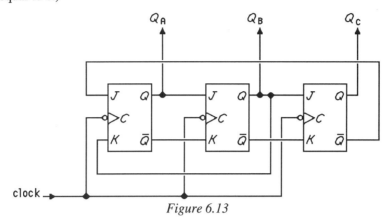

Figure 6.13

(b) What simple addition to the circuit of Figure 6.13 is required to produce a symmetrical output square wave at a frequency of one tenth of the input clock frequency?

6.8 Section 6.6.5 of the textbook contains a description of a modulo-10 counter (page 284) where the most significant bit produces a symmetrical squarewave output at a frequency of one tenth of the input clock waveform. Using the method described in Section 6.8 of the textbook, complete the state table of Table 6.3 and then design such a counter using D flip-flops and NAND gate logic. Remember that, although there are counter states which should never occur, your design must cater for the possibility of such a state occuring erroneously, and allow the correct state sequence to be rapidly re-established.

present state				next state				output
A	B	C	D	E	F	G	H	Z
0	0	0	0					0
0	0	0	1					0
0	0	1	0					0
0	0	1	1					0
0	1	0	0					0
0	1	0	1					
0	1	1	0					
0	1	1	1					
1	0	0	0					1
1	0	0	1					1
1	0	1	0					1
1	0	1	1					1
1	1	0	0					1
1	1	0	1					
1	1	1	0					
1	1	1	1					

Table 6.3

Sequential logic circuits

6.9 A sequential logic circuit is required, based on the general sequential machine of Section 6.8 of the textbook, to detect a sequence of three consecutive 1s in a serial input bit stream. The output should normally be 0, but should change to 1 on the third of the three consecutive 1s. Should more than three 1s occur consecutively, the output should remain at 1 until the next 0 is received. An example of a possible input stream and the required output is shown below.

Input (earliest bit on the left) 00101100011100101101111111011100
Output   00000000001000000000001111000100

(a) Construct the state transition diagram for the required circuit.

(b) Construct the state table in both in the state symbol version (as in Table 6.6 of the textbook) and in the binary version (as in Table 6.7 of the textbook).

(c) Construct the state assignment table for the circuit and use this table to design the required next-state logic circuit to implement the detector using AND gates, OR gates and inverters.

(d) Sketch the complete sequential circuit.

6.10 An interface circuit to be attached to a computer has stringent power switching requirements to protect the computer's operation. A power on-off switch on the interface controls the power to all integrated circuits in the interface capable of sending information to the computers internal circuits. The switch itself provides only a digital signal to a separately-powered sequential logic circuit; the output of that circuit is used to energise a relay which switches on the main interface power. The computer itself provides another input to this sequential logic circuit, a binary signal $\overline{RUN}$ which is 1 when the computer is switched on but not executing instructions, and 0 otherwise.

You are required to design this sequential logic circuit to have the following characteristics.

(i) At initial power-up of the sequential circuit (not the interface) the power to the interface must be off, whatever the position of the ON/OFF switch. At this stage, the circuit should not respond to the $\overline{RUN}$ signal becoming 1 unless the ON/OFF switch is OFF.

(ii) Once the $\overline{RUN}$ signal has become 1, changing the ON/OFF switch from OFF to ON should cause the interface power to be switched on. Similarly, providing the $\overline{RUN}$ signal remains 1, changing the ON/OFF switch from ON to OFF should result in the interface power being switched off.

(iii) Once the interface has been powered-up, a change in the $\overline{RUN}$ signal from 1 to 0 should result in the interface power remaining on irrespective of the position of the ON/OFF switch. Only when the $\overline{RUN}$ signal is once more 1 should it be possible to switch off the interface using the ON/OFF switch.

Your design should be based on the general sequential machine described in Section 6.8 of the textbook. You should first devise an appropriate state-transition diagram, then construct the state table and state assignment table to enable the next-state logic circuit to be designed. Use only NAND gates and inverters for this next-state logic. Your complete circuit should use a register with the smallest number of bits compatible with the specification of the problem.

SOLUTIONS

Question 6.1

(a)

The completed timing diagram is shown in Figure 6.14.

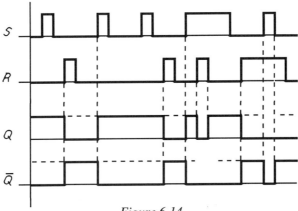

Figure 6.14

(b)

In the circuit of Figure 6.2, when inputs A and B are both 1, the Q and $\overline{Q}$ outputs of the circuit are the complement of each other, and each may be either 0 or 1, as shown in Figure 6.15.

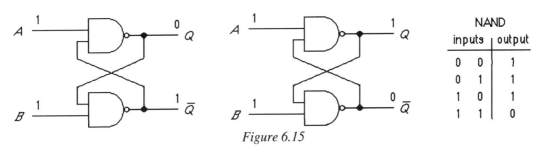

Figure 6.15

When only one of the inputs A and B becomes 0, the output of the corresponding NAND gate becomes 1, the output of the other NAND gate becoming 0. Thus a 0 on input A causes Q to become 1 (and $\overline{Q}$ to become 0), while a 0 on input B causes $\overline{Q}$ to become 1 and hence output Q to become 0.

The A input is therefore the SET (S) input to the bistable circuit while the B input is the RESET (R) input.

If both A and B become 0 at the same time, the outputs of both NAND gates become 1 and the two outputs are no longer the complement of each other.

The completed timing diagram is shown in Figure 6.16.

Sequential logic circuits

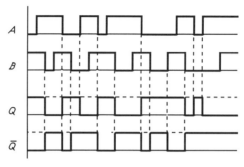

Figure 6.16

Question 6.2

(a)

By analogy with the clocked *RS* latch of Figure 6.5 of the textbook, since the inputs to the cross-coupled NAND gates of Figure 6.2 must go to 0 to perform the SET and RESET functions, the clock can conveniently be combined with set and reset signals in NAND gates, giving the circuit of Figure 6.17. Notice that the *R* and *S* inputs will need to be normally at 0, and to change to 1 to perform the setting and resetting functions when the clock also changes to 1. This circuit therefore behaves in exactly the same way as the circuit of Figure 6.5 of the textbook.

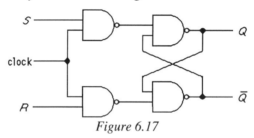

Figure 6.17

(b)

The completed timing diagram is shown in Figure 6.18. Notice the cross hatched areas. These represent times when the behaviour of the circuit is unpredictable. This is because both *R* and *S* are 1 when the clock changes from 1 to 0, and therefore the final state of the circuit will depend on delays inherent in the NAND gates, rather than on the history of the input signals.

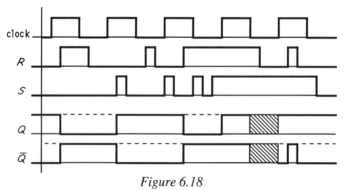

Figure 6.18

Solutions

Question 6.3

Considering Figure 6.5 and the timing diagram of Figure 6.6, since the initial output B from the Q terminal of the right-hand D flip-flop is 1, the $\overline{Q}$ output of that flip-flop must be 0. The active low CLR input of the left-hand flip-flop is therefore 0 and the Q output of this flip-flop must also be 0. This Q output is connected to the D input of the second flip-flop, so at the falling edge of the next clock pulse the Q output of the second flip-flop must change to 0. This is shown on the timing diagram of Figure 6.19 as the instant (a).

This change of state of the second flip-flop changes the CLR input of the first flip-flop to 1, removing the asynchronous clear signal, but the first flip-flop output remains 0.

Because the clock signal is not connected to the first flip-flop, as long as the asynchronous preset input PR remains at 1, no change in the state of the first flip-flop will occur. However, when the A input signal (which is connected to this PR input) falls to 0 for the first time, the Q output of the first flip-flop changes to 1 (instant (b)). At the next falling edge of the clock waveform therefore, the Q output of the second flip-flop (which is the circuit output B) will also change to 1. At the same time, the $\overline{Q}$ output of the second flip-flop changes to 0, so resetting the first flip-flop's Q output to 0 (providing the PR input of that flip-flop has by then returned to 1) (instant (c)). At the next falling edge of the clock pulse, the second flip-flop's Q output returns to 0 because of the 0 at its D input (instant (d)).

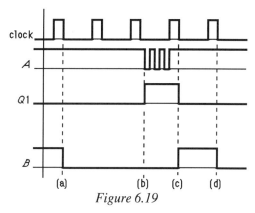

Figure 6.19

The circuit therefore, in response to a single input pulse (or to several pulses occurring very close together) produces a single output pulse of accurately defined length equal to the period of the clock signal. Sometimes, a single pulse generated at one part of a circuit can arrive at another part of the circuit with "ringing" on it, caused by stray capacitances and inductances creating a resonant circuit. The circuit of Figure 6.5 therefore "cleans up" such a pulse ready for use in other circuits.

Question 6.4

The completed table is shown in Table 6.4

Since the four flip-flops form a right-shift register, at each clock pulse the contents of the register are shifted one place to the right (the right-most bit being lost), while the signal at the D input of the left-most flip-flop is entered into the register. Since the signal at that D input is the exclusive-OR of

Sequential logic circuits

Q_C and Q_D, its value can be deduced for each state of the shift register. I have shown the value of this input in an additional column in Table 6.4 called $D1$. Whenever Q_C and Q_D are equal, the signal $D1$ is 0; whenever they are different, the value of $D1$ is 1. The value of $D1$ in any state of the register becomes the left-most bit of the register in the next state.

The pattern starts to repeat itself after the fifteenth clock pulse, and there are therefore 15 different states of the register.

number of clock pulses	Q_A	Q_B	Q_C	Q_D	$D1$
0	1	1	1	1	0
1	0	1	1	1	0
2	0	0	1	1	0
3	0	0	0	1	1
4	1	0	0	0	0
5	0	1	0	0	0
6	0	0	1	0	1
7	1	0	0	1	1
8	1	1	0	0	0
9	0	1	1	0	1
10	1	0	1	1	0
11	0	1	0	1	1
12	1	0	1	0	1
13	1	1	0	1	1
14	1	1	1	0	1
15	1	1	1	1	0

Table 6.4

Question 6.5

(a)

Assuming that the contents of all the flip-flops of Figure 6.8 are initially 0, the first clock pulse will change the state of flip-flop A from 0 to 1. Its $\overline{Q}$ output will therefore change from 1 to 0, generating a falling edge at the clock input of flip-flop B. Flip-flop B will then change state from 0 to 1, generating a falling edge at the clock input of flip-flop C. The result of this first clock pulse is therefore a change in the flip-flop contents from 000 to 111.

The second clock pulse will change the state of flip-flop A from 1 to 0, its $\overline{Q}$ output changing from 0 to 1. This rising edge at the clock input of flip-flop B will not cause any further changes of state. The second clock pulse therefore changes the contents from 111 to 011. Subsequent pulses change the contents in the sequence 011 → 101 → 001 → 110 → 010 → 100 → 000. If Q_A is considered the least significant bit and Q_B the most significant bit, this represents binary *down* counting. Writing the bits in the order $Q_C Q_B Q_A$, the sequence is 000 → 111 → 110 → 101 → 100 → 011 → 010 → 001 → 000 etc. The circuit is therefore a 3-bit natural binary ripple down counter.

Solutions

(b)

Examination of the circuit of Figure 6.9 indicates that either the Q outputs of each flip-flop or the $\overline{Q}$ outputs are gated through to the clock input of the next flip-flop by the signal X. If X is 1, the Q output is connected to the clock input, if X is 0, the $\overline{Q}$ output is connected to the clock input. The circuit therefore appears to be a natural binary ripple up/down counter, where input X controls the direction of counting; if $X = 1$, the count is up, while if X is 0, the count is down.

The completed timing diagram for the circuit is shown in Figure 6.20.

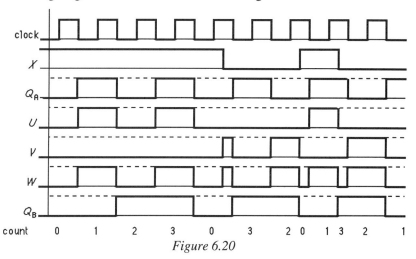

Figure 6.20

The first two stages of the counter behave as expected for the first four negative clock edges, counting up ($X = 1$) from 0 to 3 then back to 0 on the fourth edge. The fifth and sixth clock pulses are also counted correctly, the count changing from 0 to 3 to 2 (X is now 0) at the negative edges of the clock pulses. Things start to go wrong at the next change of X from 0 to 1, which initiates a change of $Q2$ which is not in synchronism with the trailing edge of the clock pulse. The count therefore changes to 0 then 1 (the count should have changed to 3) before finally reaching the correct count at the negative edge of the X pulse! The final count ends up being what it should be, but there are two intermediate spurious counts.

The counter cannot therefore be said to be working as the designer intended. The problem could be avoided by ensuring that changes in X are synchronised with the negative edge of the clock pulse.

Question 6.6

(a)   The completed table is shown in Table 6.5.

number of clock pulses	Q1	Q2	J1	K1	J2	K2
0	0	0	1	1	0	1
1	1	0	1	1	1	0
2	0	1	0	1	0	1
3	0	0	1	1	0	1

Table 6.5

105

Sequential logic circuits

Because $K1$ is always 1, $Q1$ will toggle whenever $J1$ is 1 (i.e whenever $Q2$ is 0) and will change to or remain at 0 when $J1$ is 0 (i.e when $Q2$ is 1). $Q2$ will take the previous value of $Q1$ at each clock pulse.

The completed timing diagram is shown in Figure 6.21.

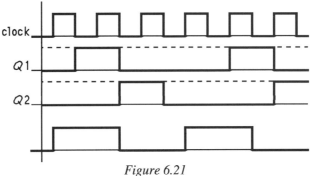

Figure 6.21

(b)

By examining the waveforms for the clock (C), $Q1$ and $Q2$, the truth table of Table 6.6 can be constructed, and hence the Karnaugh map of Figure 6.22. (The two "don't care" conditions occur because the two Qs are never 1 together.)

C	Q1	Q2	X
0	0	0	0
0	0	1	0
0	1	0	1
0	1	1	X
1	0	0	1
1	0	1	0
1	1	0	1
1	1	1	X

Table 6.6

	Q1 Q2			
C	00	01	11	10
0	0	0	X	1
1	1	0	X	1

Figure 6.22

(i) For the NOR-gate logic circuit, extract groups of 0s from the Karnaugh map, giving:

$$\overline{X} = Q2 + \overline{C}.\overline{Q1} = Q2 + \overline{C + Q1}$$

$$X = \overline{Q2 + \overline{C + Q1}}$$

(ii) For the NAND-gate circuit, extract groups of 1s, giving:

$$X = Q1 + C.\overline{Q2} = \overline{\overline{Q1}.\overline{C.\overline{Q2}}}$$

The two logic circuits are shown in Figure 6.23.

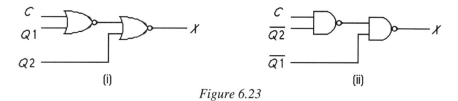

Figure 6.23

Question 6.7

(a)

Table 6.7 shows the way in which the state of the circuit changes with each input clock pulse.

number of clock pulses	Q_A	Q_B	Q_C	J_A	K_A	J_B	K_B	J_C	K_C
0	0	0	0	1	0	0	1	0	1
1	1	0	0	1	0	1	0	0	1
2	1	1	0	1	1	1	0	1	0
3	0	1	1	0	1	0	1	1	0
4	0	0	1	0	0	0	1	0	1
5	0	0	0	1	0	0	1	0	1

Table 6.7

The timing diagram is shown in Figure 6.24.

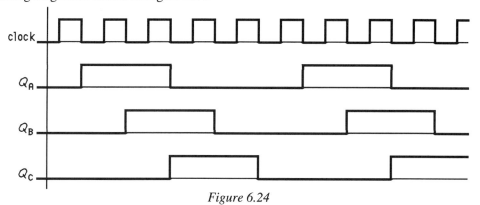

Figure 6.24

(b)

The waveforms of Figure 6.24 are at a frequency equal to one fifth of the clock frequency, but they are not symmetrical waveforms. If a waveform is generated which is exactly one half of the frequency of any of the Q outputs, triggered by the edge of a Q waveform, the result will be a symmetrical squarewave at a frequency of one tenth of the clock waveform.

Sequential logic circuits

The circuit of Figure 6.13, with the necessary addition, is shown in Figure 6.25.

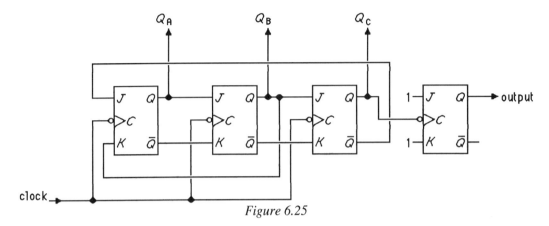

Figure 6.25

Question 6.8

The complete state table is shown in Table 6.8. There are only 10 intended counter states, and therefore there must be 6 erroneous states, which I have marked with an asterisk.

In constructing this state table, I have assumed that if the counter somehow enters one of the erroneous states, it should, on the next clock pulse, be returned to the 0000 state ready to recommence its proper counting sequence. I have kept to the simplification of using A (the most significant bit) as the symmetrical output waveform, which means that during the existence of a spurious state, the output is 1 for some and 0 for others.

present state				next state				output	
A	B	C	D	E	F	G	H	Z	
0	0	0	0	0	0	0	1	0	
0	0	0	1	0	0	1	0	0	
0	0	1	0	0	0	1	1	0	
0	0	1	1	0	1	0	0	0	
0	1	0	0	1	0	0	0	0	
0	1	0	1	0	0	0	0	0	*
0	1	1	0	0	0	0	0	0	*
0	1	1	1	0	0	0	0	0	*
1	0	0	0	1	0	0	1	1	
1	0	0	1	1	0	1	0	1	
1	0	1	0	1	0	1	1	1	
1	0	1	1	1	1	0	0	1	
1	1	0	0	0	0	0	0	1	
1	1	0	1	0	0	0	0	1	*
1	1	1	0	0	0	0	0	1	*
1	1	1	1	0	0	0	0	1	*

Table 6.8

(An equally valid alternative treatment for erroneous states would be to add a logic circuit whose output was 1 for all valid states and 0 for any erroneous state, and to use this signal to asynchronously clear all flip-flop outputs to the 0 state. This method would ensure a faster return to the valid sequence, but since this clear signal would return to 1 as soon as the counter state was once more valid, it would be possible (if some flip-flops cleared more quickly than others) for the counter to be reset to a valid state other than 0000. There would also, clearly, be the possibility of changes of state of the counter which were not in synchronism with the clock.)

The state table can also serve as the truth table for the design of the next-state logic circuit by considering A, B, C and D as the circuit inputs and E, F, G and H as the circuit outputs. The output Z will be, as I have already said, the most significant counter bit, A.

From the truth table, the Karnaugh maps for E, F, G and H can be constructed. These are shown in Figure 6.26.

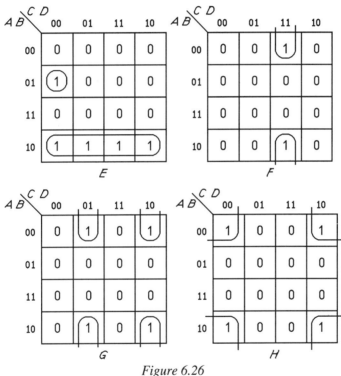

Figure 6.26

The logic functions derived from the map are:

$E = A.\bar{B} + \bar{A}.B.\bar{C}.\bar{D}$

$F = \bar{B}.C.D$

$G = \bar{B}.\bar{C}.D + \bar{B}.C.\bar{D}$

$H = \bar{B}.\bar{D}$

Sequential logic circuits

The counter will take the form shown in Figure 6.46 of the textbook (page 286). There will be no inputs other than the clock signal. The next state logic must implement the above logic functions to establish the inputs to the register ready for the next clock pulse. The output from the circuit is the most significant bit of the counter, carrying a symmetrical square wave at a frequency one-tenth of the clock frequency.

Using D flip-flops for the register the complete circuit is as shown in Figure 6.27.

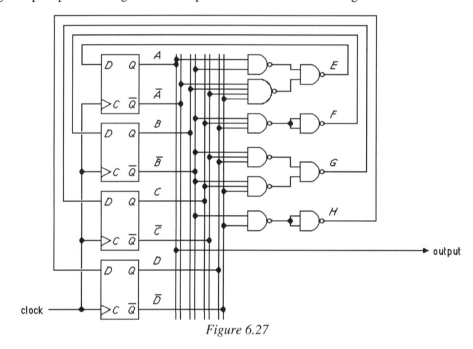

Figure 6.27

Question 6.9

(a)

One possible state transition diagram is shown in Figure 6.28.

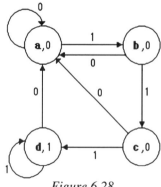

Figure 6.28

State **a** is the initial starting state, and the output is 0. A 0 in the input bit stream leaves the sequential machine in this state. A 1 in the input bit stream moves the machine to state **b**, checking for the second consecutive 1; the output is still 0. A 0 input will return the machine to state **a**, a 1 input will move the machine to state **c**, checking for the third consecutive 1. A 0 again returns the machine to state **a**, a 1 will move it to state **d** where the output becomes 1. In this fourth state, a 0 will return the machine to its initial state, while further 1s will leave the machine in the state and so maintain the 1 output.

(b)

The state table corresponding to this state transition diagram is shown in Table 6.9

present state	next state		output
	I/P = 0	I/P = 1	
a	a	b	0
b	a	c	0
c	a	d	0
d	a	d	1

Table 6.9

The state table with binary values substituted is shown in Table 6.10. The states are allocated arbitrary binary numbers (so yours may be different from mine), but each state must have a different number. For 4 states, a 2-bit register will be required.

present state	next state		output
	I/P = 0	I/P = 1	
00	00	01	0
01	00	10	0
10	00	11	0
11	00	11	1

Table 6.10

(c)

The state assignment table is shown in Table 6.11 and the Karnaugh maps for C, D and Z are shown in Figure 6.29. (See Figure 6.47 of the textbook for the significance of each variable. Z has replaced P and Q of that figure.)

A	B	U	C	D	Z
0	0	0	0	0	0
0	0	1	0	1	0
0	1	0	0	0	0
0	1	1	1	0	0
1	0	0	0	0	0
1	0	1	1	1	0
1	1	0	0	0	1
1	1	1	1	1	1

Table 6.11

Sequential logic circuits

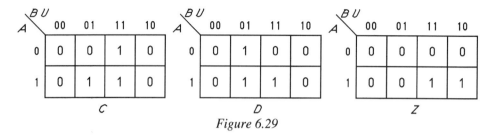

Figure 6.29

The logic expressions obtained from the Karnaugh maps are:

$$C = A.U + B.U = (A + B).U$$
$$D = A.U + \overline{B}.U = (A + \overline{B}).U$$
$$Z = A.B$$

(d)

The complete circuit of the sequential machine is shown in Figure 6.30. Note the use of one D flip-flop to ensure that changes in U do not occur at the instant the new state is being clocked into the register. This ensures that propagation delays inherent in the next-state logic cannot generate an erroneous next state. I have assumed that the clock waveform has the same frequency as the input bit stream and that the register is positive edge triggered. I have therefore included a negative edge-triggered D flip-flop so that changes in U occur one half clock cycle away from changes in the register contents.

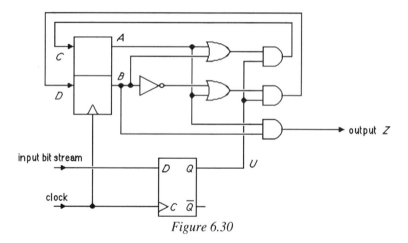

Figure 6.30

Question 6.10

Figure 6.31 is one possible state transition diagram which satisfies the conditions specified in the question. The circuit has two input signals, the ON/OFF switch S ($1 =$ ON, $0 =$ OFF) and the signal $\overline{RUN}$ from the computer. The change conditions of the state transition diagram are specified in the order S, $\overline{RUN}$. There will be one output from the circuit, the signal P which switches the interface power on and off ($P = 1$ means power on, $P = 0$ means power off).

112

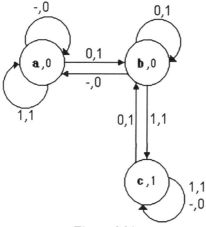

Figure 6.31

When the separate supply to the sequential circuit is first switched on, the circuit is in state **a**, where the main supply to the interface is off, whatever the position of the ON/OFF switch. (A switch-on reset circuit for the register would need to be included to ensure this starting state.)

When the $\overline{RUN}$ signal from the computer becomes 1, the circuit must change to state **b** only if the ON/OFF switch is in the OFF position. In this second state, as long as the $\overline{RUN}$ signal remains 1, the circuit waits, with the power to the interface off, for the ON/OFF switch to be moved to the ON position. Should the $\overline{RUN}$ signal become 0, the circuit returns to the initial state.

When the ON/OFF switch is moved to the ON position, the circuit moves to a state **c** in which the interface power is on. In this state, moving the ON/OFF switch to the OFF position returns the circuit to state **b**. However, if the $\overline{RUN}$ signal becomes 0, the circuit must remain in state **c** to ensure that the power supply to the interface remains on until the $\overline{RUN}$ signal again becomes 1.

The state table corresponding to this state transition diagram is shown in Table 6.12, where the next state is specified for each combination of S and $\overline{RUN}$ in each state.

present state A B	next state (C,D) (inputs $S, \overline{RUN}$)				output P
	00	01	10	11	
a	a	b	a	a	0
b	a	b	a	c	0
c	c	b	c	c	1

Table 6.12

Table 6.13 is the binary version of the state table. To simplify the logic, the unused combination of C and D (1 1) has been allocated the same output signal as state **c**, and the same next states as state **c**.

Sequential logic circuits

present state A B	next state (C,D) (inputs S, $\overline{RUN}$)				output P
	0 0	0 1	1 0	1 1	
0 0	0 0	0 1	0 0	0 0	0
0 1	0 0	0 1	0 0	1 0	0
1 0	1 0	0 1	1 0	1 0	1
1 1	1 0	0 1	1 0	1 0	1

Table 6.13

Table 6.14 is the state-assignment table derived from Table 6.13 with A, B, S and $\overline{RUN}$ considered the input variables and C, D and P considered the output variables.

A B S $\overline{RUN}$	C D P
0 0 0 0	0 0 0
0 0 0 1	0 1 0
0 0 1 0	0 0 0
0 0 1 1	0 0 0
0 1 0 0	0 0 0
0 1 0 1	0 1 0
0 1 1 0	0 0 0
0 1 1 1	1 0 0
1 0 0 0	1 0 1
1 0 0 1	0 1 1
1 0 1 0	1 0 1
1 0 1 1	1 0 1
1 1 0 0	1 0 1
1 1 0 1	0 1 1
1 1 1 0	1 0 1
1 1 1 1	1 0 1

Table 6.14

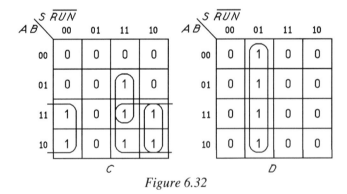

Figure 6.32

Figure 6.32 shows the Karnaugh maps for C and D derived from Table 6.14. The output P is clearly just the variable A.

The logic functions for C and D are:

$$C = A.S + A.\overline{\overline{RUN}} + B.S.\overline{RUN}$$
$$D = \overline{S}.\overline{RUN}$$

The complete circuit of the sequential machine is shown in Figure 6.33.

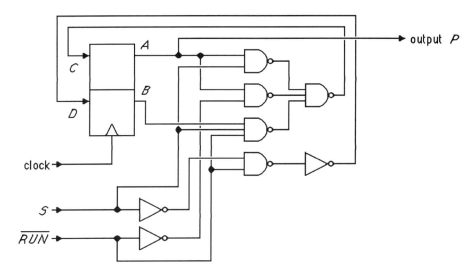

Figure 6.33

7 ANALOGUE-DIGITAL CONVERSION

QUESTIONS

7.1 **(a)** A 4-bit D-A converter, using the circuit of Figure 7.7 of the textbook, has register output voltages of +3.6 V (1) and +0.3 V (0). It has resistor values $R_1 = 2$ kΩ and $R_2 = 1$ kΩ. Assuming that the output resistances of the register are negligible in comparison with the resistances of the binary weighted network, calculate the output voltage of the converter for each of the following input codes: 0000, 1111, 1000, 0100, 1100. Comment on the linearity of the converter.

(b) What is the maximum quantisation error of the converter if it is used to produce a sawtooth waveform whose amplitude varies from 0.9 V to 10.8 V? What is the dynamic range of the converter?

7.2 **(a)** The D-A converter of Question 7.1 is to be modified to provide an output voltage of 0 V for an input code of 0000, using the circuit shown in Figure 7.1. Calculate the value of the resistor R_3.

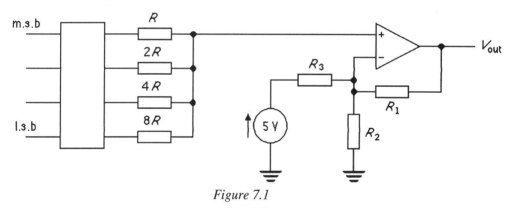

Figure 7.1

(b) What will be the output voltage for an input code of 1111?

(c) By calculating the output voltage for an input code of 1000, check whether the circuit change has affected the linearity of the converter.

7.3 **(a)** The converter of Question 7.2 is to have resistor R_3 changed in value so that the output voltage from the converter is zero when the input word is 1000. Calculate the new value of R_3.

(b) With this new value of R_3, calculate the output voltage of the converter for input words of 0000 and 1111.

(c) Are these values what you would have expected? Why has the total output voltage range increased significantly from that of the converter of Question 7.2?

117

Analogue-digital conversion

7.4 **(a)** An 8-bit *R-2R* ladder network D-A converter uses the circuit of Figure 7.12 of the textbook (but, of course, with an 8-bit register and 8 switches), with a reference voltage of +2.75 V. The required output voltage for an input code word of 00000000 is 0 V and for 11111111 is −10 V. If R_F = 40 kΩ, what must be the value of R? (You may neglect the 'on' resistance of the switches in your calculation.)

(b) The same D-A converter is now to be modified to have zero output for an input code word of 10000000 using the circuit of Figure 7.14 of the textbook, but with a positive reference voltage source of 2.75 V, rather than a negative one. The current source I_{OFF} is to be implemented using a −9.1 V source and a resistor. Calculate the required resistor value.

(c) With this new circuit, calculate the value of the output voltage for input code words of 00000000 and 11111111. Compare the total voltage ranges of the converter in its modified and unmodified form.

7.5 **(a)** A 10-bit counter ramp A-D converter, as described in Section 7.3.2 of the textbook, is required to sample adequately a signal with bandwidth 5 kHz. The signal amplitude is such as to occupy one third of the total converter input range. The converter is preceded by a sample and hold device. At what frequency must the converter clock run in order to ensure that every sample is converted to the maximum available accuracy of the converter?

(b) If the converter is now changed for a 10-bit tracking counter ramp converter, what will be the required clock frequency to give the same accuracy of conversion?

7.6 A 16-bit successive approximation converter is now used instead of a counter ramp converter to sample the signal of Question 7.5. What is the required clock frequency for this converter?

7.7 Figure 7.2 is the schematic diagram of a system containing a successive approximation A-D converter.

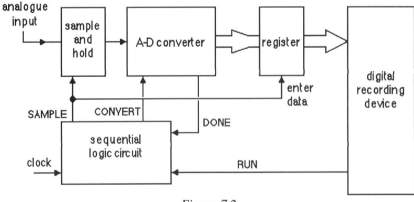

Figure 7.2

The converter is to be controlled so as to take samples at the fastest possible rate compatible with its conversion time. The A-D converter is preceded by a sample-and-hold device. The converter must be supplied with two binary signals from a sequential logic circuit which is controlling the sampling process. SAMPLE (S) is a signal which, when equal to 1 causes the sample-and-hold device to track the input voltage, and when equal to 0 causes the device to hold the current analogue voltage value. This signal also enters the previous sample from the A-D converter into a register where it is held ready for collection by the digital recording device. CONVERT (C) is a signal which, when it changes from 0 to 1 causes the A-D converter to commence conversion.

The converter has one output signal DONE (D) which changes from 0 to 1 when conversion is complete, and which changes back to 0 when the converter receives a CONVERT signal. This signal is one of the inputs to the sequential logic circuit.

There is one other signal to the system, it is a RUN (R) signal from the device which is to receive the digitized data. When this signal changes to 1, the sequential circuit must initiate taking samples, and when this signal changes to 0, the system must immediately stop sampling.

Figure 7.3 is a timing diagram of the required operation of the system.

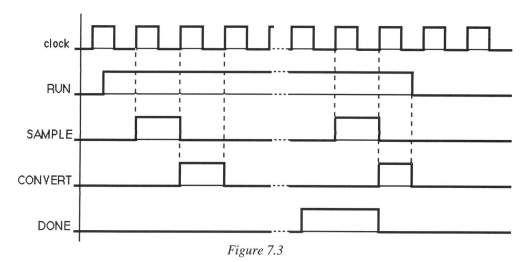

Figure 7.3

You are required to design the sequential logic circuit as an example of a general sequential machine as described in Chapter 6 of the textbook. The clock signal of Figures 7.2 and 7.3 synchronises the operation of the sequential circuit. You should construct the state transition diagram, the state tables (symbolic and binary) and the state-assignment table in order to design the next-state logic. The logic circuit should be implemented using AND gates, OR gates and inverters.

Analogue-digital conversion

SOLUTIONS

Question 7.1

(a)

With the values specified for R_1 and R_2, the closed-loop gain of the operational amplifier is

$$G = \frac{R_1 + R_2}{R_2} = 3.$$

When all input bits are 0, the equivalent circuit of the input to the operational amplifier will be as shown in Figure 7.4. Clearly, the voltage V_i, which is the input voltage to the op-amp, will be 0.3 V. The output voltage of the converter will therefore be 3×0.3 V $= 0.9$ V.

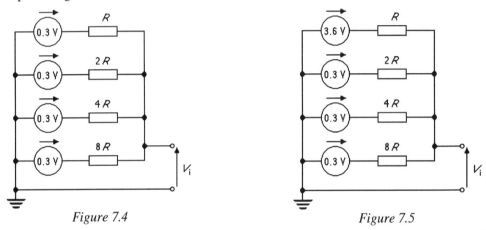

Figure 7.4   *Figure 7.5*

When all the input bits are 1, by the same reasoning, the amplifier input voltage must be 3.6 V and the converter output voltage must be 10.8 V.

When the input word is 1000, the equivalent circuit is as shown in Figure 7.5 and the input voltage to the amplifier can be calculated, using the superposition principle, as follows.

The contribution of the most significant bit will be (// stands for "in parallel with"):

$$3.6 \text{ V} \times \frac{(2R//4R//8R)}{R+(2R//4R//8R)} = 3.6 \text{ V} \times \frac{8R/7}{R+8R/7} = \frac{8}{15} \times 3.6 \text{ V}.$$

The contribution of the next most significant bit will be:

$$0.3 \text{ V} \times \frac{(R//4R//8R)}{2R+(R//4R//8R)} = 0.3 \text{ V} \times \frac{8R/11}{2R+8R/11} = \frac{4}{15} \times 0.3 \text{ V}.$$

The contribution of the third bit will be:

$$0.3 \text{ V} \times \frac{(R//2R//8R)}{4R+(R//2R//8R)} = 0.3 \text{ V} \times \frac{8R/13}{4R+8R/13} = \frac{2}{15} \times 0.3 \text{ V}.$$

The contribution of the least significant bit will be:

$$0.3\,\text{V} \times \frac{(R//2R//4R)}{8R+(R//2R//4R)} = 0.3\,\text{V} \times \frac{4R/7}{8R+4R/7} = \frac{1}{15} \times 0.3\,\text{V}.$$

Adding these contributions together gives:

$$V_i = \frac{8}{15} \times 3.6\,\text{V} + \frac{7}{15} \times 0.3\,\text{V} = 1.92\,\text{V} + 0.14\,\text{V} = 2.06\,\text{V}.$$

The output voltage from the converter is therefore $3 \times 2.06\,\text{V} = 6.18\,\text{V}$.

By similar reasoning, when the input code word is 0100, the input voltage to the amplifier will be:

$$V_i = \frac{8}{15} \times 0.3\,\text{V} + \frac{4}{15} \times 3.6\,\text{V} + \frac{2}{15} \times 0.3\,\text{V} + \frac{1}{15} \times 0.3\,\text{V} = 1.18\,\text{V}.$$

The output voltage from the converter will be $3 \times 1.18\,\text{V} = 3.54\,\text{V}$.

When the input code word is 1100, the input voltage to the amplifier will be:

$$V_i = \frac{8}{15} \times 3.6\,\text{V} + \frac{4}{15} \times 3.6\,\text{V} + \frac{2}{15} \times 0.3\,\text{V} + \frac{1}{15} \times 0.3\,\text{V} = 2.94\,\text{V}.$$

The output voltage from the converter will be $3 \times 2.94\,\text{V} = 8.82\,\text{V}$.

To check the linearity of the converter, these three voltages need to be compared with values obtained by linear interpolation between the end-point values corresponding to 0000 and 1111 inputs. Since there are 15 intervals between 0000 and 1111, the difference between the two output voltages (10.8 V – 0.9 V) must be divided by 15 to find the linear output increment. This value is therefore 0.66 V.

Table 7.1 shows the output corresponding to each input code assuming exact linearity of the converter.

input code	output voltage/V
0000	0.90
0001	1.56
0010	2.22
0011	2.88
0100	3.54
0101	4.20
0110	4.86
0111	5.52
1000	6.18
1001	6.84
1010	7.50
1011	8.16
1100	8.82
1101	9.48
1110	10.14
1111	10.80

Table 7.1

Comparing the calculated output values with the assumed linear values shows that the converter is linear.

(b)

The maximum quantisation error is equal to the output interval and is therefore 0.66 V.

The dynamic range is $20\log_{10}(2^n - 1) = 20\log_{10} 15 = 23.5 \text{ dB}$.

Question 7.2

(a)

With the circuit of Figure 7.1, the voltage at the non-inverting input of the amplifier is +0.3 V when the input code is 0000 (see the solution to Question 7.1 part (a)). The 5 V supply is therefore required to provide an equal voltage at the inverting input of the amplifier so as to balance the voltage at the non-inverting input and so allow the amplifier output to be zero. (Remember, the op-amp output is an amplified version of the *difference* between the two input voltages.)

The output of the amplifier, together with the feedback potential divider R_1 and R_2 can be represented by its Thévenin equivalent circuit, as shown in Figure 7.6, where $V_T = V_o/3$ and R_T is the parallel combination of R_1 and R_2 and is therefore equal to $2/3$ kΩ.

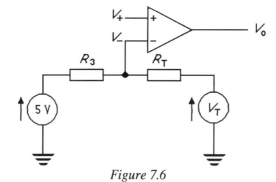

Figure 7.6

Because the current flowing into the amplifier input terminals is negligibly small, the equivalent resistance of the binary weighted resistor network plays no part in the calculations.

The required value of R_3 can be calculated from Figure 7.6 as follows.

As already stated, when the input code word is 0000 the voltage at the non-inverting input to the amplifier is +0.3 V. Assuming the amplifier to be ideal, the voltage at the inverting input must also be +0.3 V while the amplifier output voltage (and hence V_T) is zero. So,

$$\frac{(5-0.3) \text{ V}}{R_3} = \frac{0.3 \text{ V} - V_T}{R_T} = \frac{0.3 \text{ V}}{2/3 \text{ k}\Omega}$$

$$R_3 = \frac{4.7 \text{ V}}{0.3 \text{ V}} \times 2/3 \text{ k}\Omega = 10.4 \text{ k}\Omega.$$

Solutions

(b)

When the input code word is 1111, the voltage at the non-inverting input is +3.6 V, so the voltage at the inverting input must also be +3.6 V. The output voltage can therefore be calculated (again using Figure 7.6) as follows:

$$\frac{(5-3.6)\text{ V}}{R_3} = \frac{3.6\text{ V} - V_o/3}{\tfrac{2}{3}\text{ k}\Omega}$$

$$\frac{1.4\text{ V} \times \tfrac{2}{3}\text{ k}\Omega}{10.4\text{ k}\Omega} = 3.6\text{ V} - V_o/3$$

$$V_o = 3 \times \left(3.6\text{ V} - \frac{2.8}{31.2}\text{ V}\right) \approx 10.5\text{ V}.$$

(c)

When the input code word is 1000, the voltage at the non-inverting input of the amplifier is +2.06 V (see solution to Question 7.1 part (a)). The voltage at the inverting input must therefore also be +2.06 V. The output voltage is therefore given by:

$$\frac{(5-2.06)\text{ V}}{R_3} = \frac{2.06\text{ V} - V_o/3}{\tfrac{2}{3}\text{ k}\Omega}$$

$$\frac{2.94\text{ V} \times \tfrac{2}{3}\text{ k}\Omega}{10.4\text{ k}\Omega} = 2.06\text{ V} - V_o/3$$

$$V_o = 3 \times \left(2.06\text{ V} - \frac{5.88}{31.2}\text{ V}\right) \approx 5.61\text{ V}.$$

This value must be compared with the assumed linear value for an input of 1000 which, by analogy with the calculation in Question 7.1 part (a) will be:

$$\frac{(10.5\text{ V} - 0\text{ V})}{15} \times 8 = 5.6\text{ V}.$$

Within the limits of accuracy of the calculation, the converter appears linear.

Question 7.3

(a)

Since the voltage at both amplifier input terminals is +2.06 V when the input word is 1000, the new value of R_3 can be calculated using the circuit of Figure 7.6 as follows.

$$\frac{(5-2.06)\text{ V}}{R_3} = \frac{(2.06-0)\text{ V}}{\tfrac{2}{3}\text{ k}\Omega}$$

$$R_3 = \frac{2.94\text{ V}}{2.06\text{ V}} \times \tfrac{2}{3}\text{ k}\Omega = 0.95\text{ k}\Omega.$$

Analogue-digital conversion

(b)

For an input word of 0000, the amplifier input terminal voltages are both +0.3 V.

$$\frac{(5-0.3)\text{ V}}{0.95\text{ k}\Omega} = \frac{0.3\text{ V} - V_0/3}{\tfrac{2}{3}\text{ k}\Omega}$$

$$\frac{4.7\text{ V} \times \tfrac{2}{3}\text{ k}\Omega}{0.95\text{ k}\Omega} = 0.3\text{ V} - V_0/3$$

$$V_0 = 3 \times (0.3\text{ V} - 3.3\text{ V}) = -9\text{ V}.$$

For an input word of 1111, the amplifier input voltages are both +3.6 V.

$$\frac{(5-3.6)\text{ V}}{0.95\text{ k}\Omega} = \frac{3.6\text{ V} - V_0/3}{\tfrac{2}{3}\text{ k}\Omega}$$

$$\frac{1.4\text{ V} \times \tfrac{2}{3}\text{ k}\Omega}{0.95\text{ k}\Omega} = 3.6\text{ V} - V_0/3$$

$$V_0 = 3 \times (3.6\text{ V} - 0.98\text{ V}) \approx 7.9\text{ V}.$$

(c)

If you have been regarding the addition of the +5 V supply and R_3 as introducing just a d.c. shift in the converter output, you are probably surprised to find that the total output voltage range has increased from the 9.9 V (10.8 V – 0.9 V) of Question 7.1(a) to 16.9 V in part (b) of this question. The reason for this increase is that the voltage gain of the op-amp has also been increased by the introduction of R_3. The gain has been increased because, looking from the amplifier output into the feedback network, R_3 is effectively in parallel with R_2, so reducing its value. With an R_3 of 0.95 kΩ, the effective value of R_2 has been reduced to a little less than 500 Ω. This reduction in R_2 reduces the feedback fraction β, in this case to about one fifth, and so increases the closed-loop gain of the amplifier to about 5.

Question 7.4

(a)

When the input code word is 11111111, the current flowing in R_F will be $255/256 \times I$. Since the output voltage is required to be 10 V, the required value of the current I will be given by:

$$\frac{255}{256} I \times R_F = 10\text{ V} \quad \text{or} \quad I = \frac{2560\text{ V}}{255 \times 40\text{ k}\Omega} = 0.25\text{ mA}.$$

Now $I = V_{\text{ref}}/R$ so $R = V_{\text{ref}}/I = (2.75\text{ V})/(0.25\text{ mA}) = 11\text{ k}\Omega$.

(b)

When the input code word is 10000000, the current I_s flowing in the feedback resistor in the unmodified converter is $I/2 = 0.125$ mA. To reduce this current to zero and hence obtain zero output voltage, an offset current of 0.125 mA must be extracted from the virtual earth point. This can be established using the –9.1 V source and a resistor of value 9.1 V ÷ 0.125 mA ≈ 73 kΩ.

(c)

When the input code word is 00000000 the input current from the R-$2R$ ladder network will be zero, so the current I_s in the feedback resistor will be just the offset current I_{OFF} (which is 0.125 mA) and the output voltage will be + 0.125 mA × 40 kΩ = +5 V.

When the input code word is 11111111 the input current from the ladder network will be $255/256 \times I$ and the total current through the feedback resistor will be $(255/256 \times I) - I_{OFF}$. The output voltage is therefore:

$$V_o = -[255/256\, I - I_{OFF}] R_F$$
$$\approx -[0.25 \text{ mA} - 0.125 \text{ mA}] \times 40 \text{ k}\Omega$$
$$= -5 \text{ V}.$$

The total output voltage range of the converter is 10 V in both the modified and unmodified form. The addition of R_3 has not affected the gain of the amplifier, merely introduced a d.c. offset to the output voltage.

[The reason why the gain has not been affected in this converter (as it was in the converter of Question 7.3) is that this time the non-inverting input of the amplifier is connected to 0 V, so that the inverting input is a virtual earth point. The addition of extra resistance between the virtual earth and 0 V cannot affect the overall amplifier gain from signal input to signal output.]

Question 7.5

(a)

To sample a 5 kHz bandwidth signal adequately requires 10 000 samples per second. The total conversion time must therefore not exceed 0.1 ms.

A 10-bit converter requires 1024 clock pulses to change the output through its full range. If the counter is reset to zero at the start of each conversion, since the question does not specify *where* in the range of outputs the signal voltage lies, the maximum permitted conversion time must provide time for the full 1024 clock pulses.

The required clock frequency is therefore 10.24 MHz.

(b)

With a tracking counter ramp converter, the maximum possible change in output between any two successive samples is one third of 1024 (since the input signal amplitude occupies only one third of the total input range) or 342 clock pulses.

Analogue-digital conversion

The required clock frequency is therefore 3.42 MHz.

[This is a very conservative estimate of the required clock frequency. If the bandwidth of the signal is 5 kHz, since the amplitudes of the higher frequency components fall as the frequency increases, the likelihood of a 5 kHz component at an amplitude equal to one third of the total available range is extremely remote. A significantly lower frequency would therefore almost certainly be adequate.]

Question 7.6

A 16-bit successive approximation converter requires 17 clock pulses to achieve each conversion (one CLEAR pulse plus one pulse for each bit).

The required clock frequency is therefore 170 kHz.

Question 7.7

Figure 7.7 shows one possible state transition diagram for the sequential circuit. The input variables are listed in the order RUN (R), DONE (D) while the output variables are listed in the order SAMPLE (S), CONVERT (C).

Figure 7.8 shows the link between the state transition diagram and the timing diagram.

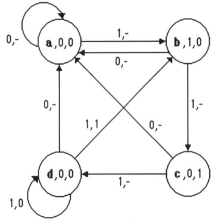

Figure 7.7

The state table corresponding to this state transition diagram is shown in Table 7.2.

present state (W,X)	next state (Y,Z) inputs (R,D)				outputs (S,C)
	00	01	10	11	
a	a	a	b	b	0 0
b	a	a	c	c	1 0
c	a	a	d	d	0 1
d	a	a	d	b	0 0

Table 7.2

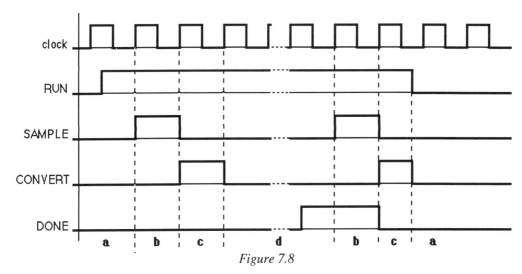

Figure 7.8

The state table with binary values is shown in Table 7.3

present state (W,X)	next state (Y,Z) inputs (R,D)				outputs (S,C)
	00	01	10	11	
0 0	0 0	0 0	0 1	0 1	0 0
0 1	0 0	0 0	1 0	1 0	1 0
1 0	0 0	0 0	1 1	1 1	0 1
1 1	0 0	0 0	1 1	0 1	0 0

Table 7.3

The state-assignment table is shown in Table 7.4

W	X	R	D	Y	Z	S	C
0	0	0	0	0	0	0	0
0	0	0	1	0	0	0	0
0	0	1	0	0	1	0	0
0	0	1	1	0	1	0	0
0	1	0	0	0	0	1	0
0	1	0	1	0	0	1	0
0	1	1	0	1	0	1	0
0	1	1	1	1	0	1	0
1	0	0	0	0	0	0	1
1	0	0	1	0	0	0	1
1	0	1	0	1	1	0	1
1	0	1	1	1	1	0	1
1	1	0	0	0	0	0	0
1	1	0	1	0	0	0	0
1	1	1	0	1	1	0	0
1	1	1	1	0	1	0	0

Table 7.4

Analogue-digital conversion

The Karnaugh maps obtained from Table 7.4 are shown in Figure 7.9 and the logic functions derived from these maps are:

$$Y = \overline{W}.X.R + W.\overline{X}.R + W.R.\overline{D}$$
$$Z = \overline{X}.R + W.R$$
$$S = \overline{W}.X$$
$$C = W.\overline{X}$$

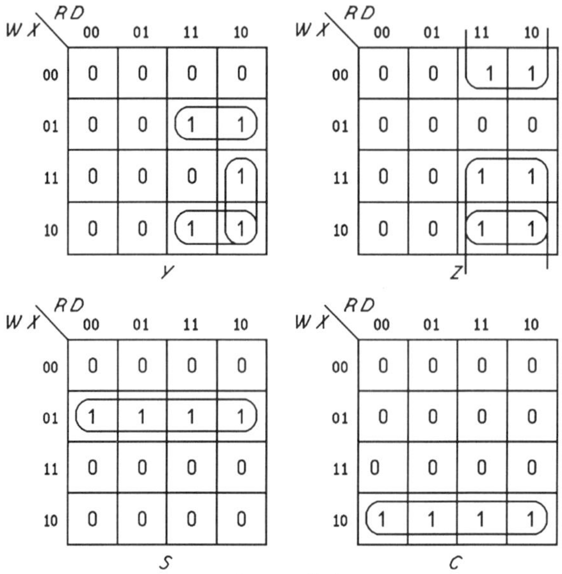

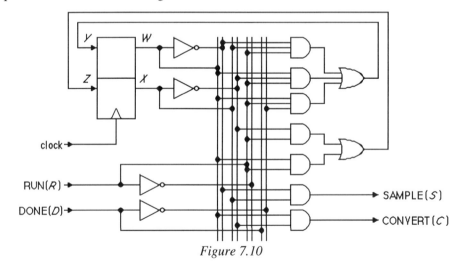

Figure 7.9

The sequential circuit is shown in Figure 7.10

Figure 7.10

8 MEMORIES, MICROPROCESSORS AND MICROCONTROLLERS

QUESTIONS

8.1 Figure 8.1 shows the structure of a memory for a microprocessor. All RAM chips are identical.

    (a)   (i)   What is the total RAM capacity?

            (ii)   What is the capacity of ROM A?

            (iii) What is the size of ROM B?

            (iv) What is the size of the EPROM?

    (b)   Assuming that, where not all address lines are connected to a chip, only the low-order available bits are connected, what is the range of addresses occupied by:

            (i)   ROM A?

            (ii) RAM?

            (ii) ROM B?

            (iv) EPROM?

(You may quote addresses using either denary or hexadecimal notation.)

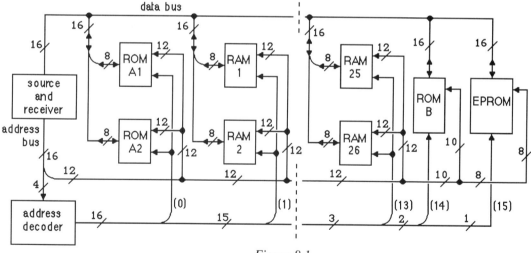

Figure 8.1

Memories, microprocessors and microcontrollers

8.2 A section of program for a microprocessor consists of the instructions shown below.

address	instruction
32, 33	load the accumulator with the contents of location 80
34, 35	branch if the accumulator is zero to location 47
36	decrement the accumulator contents
37, 38	store the accumulator contents in location 80
39, 40	load the accumulator with the contents of location 81
41, 42	add to the accumulator the contents of location 81
43, 44	store the accumulator contents in location 81
45, 46	jump to location 32
47, 48	load the accumulator with the contents of location 81
⋮	
80	5
81	1

(a) Assuming that, prior to execution of the program segment, locations 80 and 81 contain the values shown in the table, what will be the accumulator contents after execution of the instruction in locations 47 and 48?

[**Hint:** list the contents of locations 80 and 81 each time the instructions are executed.]

(b) If the initial contents of location 80 had been n and the contents of location 81 had been x, what would be the contents of the accumulator after execution of the instruction in locations 47 and 48?

8.3 An MC68HC05B6 is to be set up so that port A and pins 4 to 7 of port B form a 12-bit data input port, while pins 0 to 3 of port B and port C form a 12-bit data output port.

(a) What data must be placed in which storage locations to achieve this?

(b) Once set up, what instructions are required to (i) collect data from the input port and (ii) write data to the output port?

(c) What is the result of running the following program segment, assuming the ports have been set up as described?

addresses	instructions
32, 33	load the accumulator with 0
34, 35	store the accumulator contents in location 1
36, 37	load the accumulator with 255_{10}
38, 39	store the accumulator contents in location 2
40, 41	load the accumulator with the contents of location 0
42, 43	compare the accumulator contents with 0
44, 45	branch to 48 if they are equal
46, 47	jump to location 40

continued...

addresses	instructions
48, 49	load the accumulator with the contents of location 1
50, 51	compare the accumulator contents with 0
52, 53	branch to 56 if they are equal
54, 55	jump to location 40
56, 57	load the accumulator with 15
58, 59	store the accumulator contents in location 1
60, 61	load the accumulator with 0
62, 63	store the accumulator contents in location 2
64	continue

8.4 **(a)** The A-D converter on the MC68HC05B6 is required to measure analogue voltages so that an input of +3 V generates a digital value of 10000000_2 and an input of +3.8 V generates a value of 10110010_2. What must be connected to terminals VRH and VRL of the microcontroller?

(b) What short section of program is required to select channel 4 on the A-D converter multiplexer, wait for conversion to be complete and then read in the converted value to memory location 80_{10}. Assume that the program segment starts at location 32_{10}.

(c) Give the denary (or hexadecimal) code for the instructions which select the channel, start the conversion and test whether conversion is complete. (The two-byte instruction "load the accumulator with the value n" has the op-code 166 as the first byte and the number n as the second byte.)

(d) The MC68HC05B6 is now required to sample the signals on all 8 analogue input channels in turn. The processor is supplied with a quartz crystal which fixes the oscillator frequency at 4 MHz. Assuming that the instruction "load the accumulator with a constant" takes 2 clock cycles, that the instruction "store the accumulator contents in location n" takes 4 clock cycles and that the instruction "branch if bit n is set" takes 5 clock cycles, what is the maximum permitted frequency component of the input analogue signals?

8.5 A step waveform has sample values 0 (prior to the step) and 200_{10} (after the step). A digital filter defined by the equation

$$y_n = x_n - x_{n-1} + 0.8\, y_{n-1}$$

is applied to this sampled step waveform.

(a) Calculate the output values and sketch the resulting output waveform.

(b) What sort of filter is this?

(c) If the sample time is 0.5 ms, what is the time constant of the filter?

(d) With the same sample time, what is the cut-off frequency of the filter?

Memories, microprocessors and microcontrollers

SOLUTIONS

Question 8.1

(a)

(i) Each RAM chip must have capacity 4096 (12 input address lines) by 8 (8 data lines). Each pair of RAM chips therefore has a capacity of 4096 × 16. and the total RAM capacity will be 52 K × 16 (13 pairs of chips).

(ii) By similar reasoning, the capacity of ROM A will be 4 K × 16.

(iii) The size of ROM B must be 1 K × 16 (since it has 10 address lines).

(iv) The size of the EPROM must be 256 × 16.

(b)

(i) ROM A is addressed by bit 0 from the address decoder, which is generated by an input to the decoder of 0000, plus the 12 low-order address lines. It therefore occupies addresses between 0000000000000000_2 and 0000111111111111_2. This corresponds to the range 0 to 4095_{10} or 0 to $0FFF_{16}$.

(ii) RAM is addressed by codes 0001 to 1101 on the four high-order address bits, plus the 12 low-order address bits. It therefore occupies addresses between 0001000000000000_2 and 1101111111111111_2 which corresponds with 4096_{10} to 57343_{10} or 1000_{16} to $DFFF_{16}$.

(iii) ROM B is addressed by the code 1110 on the four high-order address bits, plus the ten low-order address bits. Its address range is therefore from $1110xx0000000000_2$ to $1110xx1111111111_2$ (the xs mean that the value of that bit doesn't matter). The ROM, although only having a capacity of 1 K 16-bit words, actually occupies the address range 1110000000000000_2 (when the two unused bits are both 0) to 1110111111111111_2 (when the two unused bits are both 1). This range corresponds with the range 57344_{10} to 61439_{10} or $E000_{16}$ to $EFFF_{16}$.

(iv) The EPROM is addressed by code 1111 on the four high-order address bits plus the eight low-order address bits. Its address range is therefore from $1111xxxx00000000_2$ to $1111xxxx11111111_2$. This corresponds to the range 1111000000000000_2 (when the unused bits are all 0) to 1111111111111111_2 (when the unused bits are all 1). This corresponds with 6144_{10} to 65535_{10} and $F000_{16}$ to $FFFF_{16}$.

Question 8.2

(a)

Each time the program finds the contents of location 80 non-zero it decrements the contents of location 80, and doubles the contents of location 81 before again examining the contents of location 80. Since the initial contents of location 80 were 5, the contents of locations 80 and 81 will change as follows:

Solutions

	start	contents after:					
address	contents	run 1	run 2	run 3	run 4	run 5	run 6
80	5	4	3	2	1	0	0
81	1	2	4	8	16	32	32

The final value contained in location 81 is 32 or 2^5.

(b)

If the original contents of location 80 had been n, then the contents of location 81 would be doubled n times. If the original contents of location 81 had been x, then the final contents would be $2^n x$.

Question 8.3

(a)

To make all the bits of port A input lines the value 00000000_2 must be placed in the port A data direction register (DDR), i.e. in memory location 4_{10}.

To make the four high-order bits of port B input lines, and the four low-order bits output lines, the value 00001111_2 must be placed in the port B DDR, i.e. in memory location 5_{10}.

To make all the bits of port C output lines, the value 11111111_2 must be placed in memory location 6_{10} (the port C DDR).

(b)

(i) To collect data from the input port the processor must read the contents of memory locations 0 and 1, which are the port A and port B data registers.

(ii) To write data to the output port the processor must store data in memory locations 1 and 2 which are the port B and port C data registers.

(c)

The program first puts data into port B and port C data registers to send out the 12-bit pattern 000011111111_2 (255_{10}). It then reads the contents of location 0 (the port A data register) and tests for zero. If the contents are zero, the program proceeds to read and test the contents of the port B data register. If that too is zero, the program writes the pattern 00001111_2 to port B and 00000000_2 to port C. Hence if the data on the 12-bit input port is zero, the 12-bit output port data changes to 111100000000_2 (or 3840_{10}).

If either of the tests of the contents of locations 0 and 1 show the input data to be non-zero, the output port continues to carry the data 000011111111_2 and the program branches back to the instruction in locations 40 and 41 to continue testing the input ports.

Memories, microprocessors and microcontrollers

Question 8.4

(a)

The difference between the digital values at analogue input voltages of +3 V and +3.8 V is $110010_2 = 50_{10}$. The input voltage difference is 0.8 V. The required A-D converter sensitivity is therefore 16 mV per bit.

The output code 00000000_2 therefore corresponds to an input of $3\text{ V} - (128 \times 16\text{ mV}) = +0.952$ V.

The output code 11111111_2 corresponds to an input voltage of $3\text{ V} + (127 \times 16\text{ mV}) = +5.032$ V.

Hence, +0.952 V must be supplied to pin VRL and +5.032 V to pin VRH.

(b)

The program segment will be:

addresses	instructions
32, 33	load the accumulator with 36_{10}
34, 35	store the accumulator contents in location 9
36, 37, 38	branch to location 41 if bit 7 is set in location 9
39, 40	jump to location 36
41	continue processing

The instructions in locations 32, 33 and 34, 35 select channel 4 and start the conversion.

The instruction in locations 36, 37 and 38 checks whether bit 7 has been set in location 9 by the completion of the conversion process.

(c)

The binary, denary and hexadecimal codes for the instructions are shown in Table 8.1

address	binary	decimal	hex
32	10100110	166	A6
33	00100100	36	24
34	10110111	183	B7
35	00001001	9	09
36	00001110	14	0E
37	00001001	9	09
38	00000010	2	02

Table 8.1

(d)

If the oscillator frequency is 4 MHz, the processor clock frequency is 2 MHz.

Selecting a channel and starting conversion uses one **load a constant** instruction and one **store** instruction, and hence requires 6 clock cycles and takes 3 µs.

The conversion process requires 32 clock cycles and so takes 16 µs.

Checking for conversion complete requires 5 clock cycles and so takes 2.5 µs.

Assuming that the process of collecting the data can overlap with the next conversion process, the total time for obtaining one sample on one channel is therefore 21.5 µs.

The total time required to sample all 8 channels is therefore 172 µs. The maximum sample rate on each channel is therefore 1/(172 µs) = 5814 samples per second.

The maximum permitted frequency component of each analogue input signal is therefore half this sample rate, or approximately 2.9 kHz.

Question 8.5

(a)

Assuming that the step input occurs some time between samples 2 and 3 being taken, Table 8.2 shows the values generated at the filter output (x_{n-1} does not have a valid value until sample 2 has been taken). These values are plotted in Figure 8.2

n	x_n	$x_n - x_{n-1}$	y_{n-1}	$0.8\, y_{n-1}$	$y_n = x_n - x_{n-1} - 0.8\, y_{n-1}$
0	0	–	0	0	–
1	0	–	0	0	–
2	0	0	0	0	0
3	200	200	0	0	200
4	200	0	200	160	160
5	200	0	160	128	128
6	200	0	128	102.4	102.4
7	200	0	102.4	81.9	81.9
8	200	0	81.9	65.5	65.5
9	200	0	65.5	52.4	52.4
10	200	0	52.4	41.9	41.9
11	200	0	41.9	33.6	33.6
12	200	0	33.6	26.8	26.8
13	200	0	26.8	21.5	21.5
14	200	0	21.5	17.2	17.2
15	200	0	17.2	13.7	13.7
16	200	0	13.7	11.0	11.0
17	200	0	11.0	8.8	8.8
18	200	0	8.8	7.0	7.0

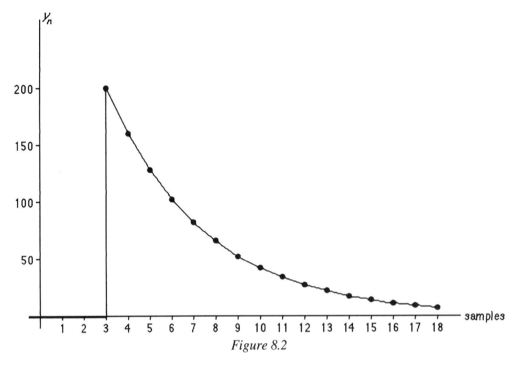

Figure 8.2

(b)

The shape of the step response of the filter indicates that it is a first-order high-pass filter.

(c)

The time constant can be estimated from the step response, since the output will fall through 95% of its total decay in a time equal to 3 time constants. 95% of 200 is 190, so the time taken for the output to reach the value 10 will be 3 time constants.

If the sampling interval is 0.5 ms, the time to reach 10 V is about $(16-3) \times 0.5$ ms = 6.5 ms.

The time constant is therefore $6.5 \div 3 \approx 2.2$ ms.

(d)

The cut-off frequency of the filter is given by $\omega_c \tau = 1$ where τ is the time constant, so

$$\omega_c = 1000/2.2 \text{ rad s}^{-1}$$

and

$$f_c = \frac{1000}{2\pi \times 2.2} \text{ Hz}$$
$$= 72 \text{ Hz}.$$

9 DIODES AND TRANSISTORS

QUESTIONS

9.1 **(a)** A thin cylindrical slice of silicon has diameter 5 cm and thickness 0.4 mm. The resistance measured between the flat faces is 100 Ω. What is the conductivity of the silicon?

(b) The temperature of the silicon slice is increased by 25 °C. Assuming that the conductivity of silicon increases by 8% per °C, what will be the new slice resistance?

9.2 Silicon slices dimensionally identical to those of Question 9.1 are doped with:

(a) acceptors to a density of 10^{22} m^{-3};

(b) donors to a density of 10^{21} m^{-3};

(c) donors to a density of 5×10^{20} m^{-3} and acceptors to a density of 10^{21} m^{-3}.

Calculate the resistance of each slice measured between the flat faces.

9.3 **(a)** A piece of n-type silicon has a doping density of 3×10^{20} m^{-3}. What is the minority carrier density p_o at 25 °C?

(b) What is the minority carrier density at 25 °C in a piece of p-type silicon having doping density 5×10^{21} m^{-3}?

(c) What will be the minority carrier densities of the two pieces of silicon at a temperature of 100 °C?

(d) For each piece of silicon, calculate the temperature at which the minority carrier density becomes 10% of the majority carrier density.

9.4 **(a)** The p-region of a pn junction is doped with boron to a density of 3×10^{21} atoms m^{-3}. The junction is forward biassed by 0.5 V. What is the density of electrons in the p-region next to the depletion region if $T = 25$ °C and $K = 40$ V^{-1}?

(b) If the temperature increases by 20 °C, what is the new value of the electron density?

(c) The same pn junction is now reverse biassed with a voltage of –0.15 V. What is the electron density in the p-region next to the transition region?

(d) If the temperature of the pn junction in (c) increases by 20 °C, what is the new value of the electron density?

Diodes and transistors

9.5 **(a)** A silicon diode has a value of I_s of 3×10^{-13} A and a K of 40 V^{-1}. Calculate the current flowing in the diode at forward bias voltages of (i) 0.5 V, (ii) 0.6 V, (iii) 0.7 V and (iv) 0.8 V.

(b) Re-calculate the values of current for a diode having the same value of I_s but a value of K of 35 V^{-1}.

(c) For the diodes of parts (a) and (b), calculate the reverse bias voltage at which the amplitude of the reverse current becomes 99% of the saturation current I_s.

9.6 In the diode equation $I_D = I_s\left(e^{KV_D} - 1\right)$, the theoretical value of K is q/kT where q is the charge on an electron, k is Boltzmann's constant (1.38×10^{-23} J K^{-1}) and T is the temperature in kelvin (see the margin note on page 392 of the textbook).

(a) Calculate the value of K at (i) 25 °C and (ii) 50 °C.

(b) For a diode with $I_s = 3 \times 10^{-13}$ A at 25 °C, calculate the current flowing when the forward bias is 0.6 V (i) at 25 °C and (ii) at 50 °C.

(Remember, I_s will also be affected by temperature changes.)

9.7 A bipolar transistor having $\beta = 300$, $V_A = 120$ V and $I_{SE} = 2 \times 10^{-13}$ A at $V_{CE} = 10$ V is connected in common-emitter configuration.

(a) If $V_{CE} = 5$ V and $V_{BE} = 0.65$ V, calculate the values of I_C, I_E and I_B assuming that $K = 40$ V^{-1}.

(b) What is the output resistance of the transistor at this operating point?

SOLUTIONS

Question 9.1

(a)

The equation for the resistance of a piece of silicon is $R = l/\sigma a$ where l is the length, a is the cross-sectional area and σ is the conductivity of the silicon. Hence:

$$\sigma = \frac{1}{R} \times \frac{l}{a} = 0.01 \times \frac{0.4 \times 10^{-3}}{\pi \times (5 \times 10^{-2})^2 / 4} \text{ S m}^{-1}$$

$$= 0.01 \times 0.204 \text{ S m}^{-1}$$

$$= 2.04 \times 10^{-3} \text{ S m}^{-1}$$

(b)

If the temperature of the silicon is increased by 25 °C, the conductivity will rise by a factor of $(1.08)^{25} = 6.85$, so the conductivity rises to

$$\sigma = 2.04 \times 10^{-3} \times 6.85 \text{ S m}^{-1}$$

$$= 14.0 \times 10^{-3} \text{ S m}^{-1}$$

Question 9.2

(a)

Conductivity $= \sigma_p = N_a q \mu_p = 10^{22} \times 1.6 \times 10^{-19} \times 0.045 \text{ S m}^{-1} = 72 \text{ S m}^{-1}$

Hence the resistance $= l/\sigma_p a = \frac{1}{72} \times 0.204 \text{ }\Omega = 2.8 \times 10^{-3} \text{ }\Omega$

(b)

Conductivity $= \sigma_a = N_p q \mu_a = 10^{21} \times 1.6 \times 10^{-19} \times 0.15 \text{ S m}^{-1} = 24 \text{ S m}^{-1}$

Hence, the resistance $= l/\sigma_a a = \frac{1}{24} \times 0.204 \text{ }\Omega = 8.5 \times 10^{-3} \text{ }\Omega$

(c)

Since $N_a = 10^{21} \text{ m}^{-3}$ and $N_d = 5 \times 10^{20} \text{ m}^{-3}$,

Resulting acceptor density $(N_a - N_d) = 5 \times 10^{20} \text{ m}^{-3}$.

Hence, $\sigma = 5 \times 10^{20} \times 1.6 \times 10^{-19} \times 0.045 \text{ S m}^{-1} = 3.6 \text{ S m}^{-1}$

and the resistance $= \frac{1}{3.6} \times 0.204 \text{ }\Omega = 56 \times 10^{-3} \text{ }\Omega$

Question 9.3

(a)

Using the equation $p_0 n_0 = n_i^2$ and the value of $n_i = 1.5 \times 10^{16}$ at 25 °C,

$$p_0 \times 3 \times 10^{20} = (1.5 \times 10^{16})^2$$

$$p_0 = 0.75 \times 10^{12} \text{ m}^{-3}$$

(b)

Similarly, $\quad n_0 \times 5 \times 10^{21} = (1.5 \times 10^{16})^2$

so, $\quad n_0 = 0.45 \times 10^{11} \text{ m}^{-3}$

(c)

At 100 °C,

$$n_i^2 = \left(1.5 \times 10^{16} \times (1.08)^{75}\right)^2 \text{ m}^{-6}$$

$$= 2.25 \times 10^{32} \times (1.08)^{150} \text{ m}^{-6}$$

$$= 2.25 \times 10^{32} \times 10.3 \times 10^{4} \text{ m}^{-6}$$

so n_i^2 is increased by a factor 10.3×10^4 and so the minority carrier density is also increased by this factor.

In the n-type silicon, $p_0 = 0.75 \times 10^{12} \times 10.3 \times 10^4 \text{ m}^{-3} = 7.7 \times 10^{16} \text{ m}^{-3}$

For the p-type silicon, $n_0 = 0.45 \times 10^{11} \times 10.3 \times 10^4 \text{ m}^{-3} = 4.6 \times 10^{15} \text{ m}^{-3}$

(d)

If the temperature at which the minority carrier density becomes equal to 10% of the majority carrier density is T °C, then the temperature increase from 25 °C is $(T-25)$ °C.

The increase in n_i^2 will be $\left[(1.08)^{(T-25)}\right]^2 = (1.08)^{(2T-50)}$

For the n-type silicon, $n_0 = 3 \times 10^{20}$ so $p_0 = \dfrac{2.25 \times 10^{32} \times (1.08)^{(2T-50)}}{3 \times 10^{20}}$

At T °C, $p_0 = 0.1 \times n_0 = 3 \times 10^{19}$ so, $(1.08)^{(2T-50)} = \dfrac{3 \times 10^{19} \times 3 \times 10^{20}}{2.25 \times 10^{32}} = 4 \times 10^7$

Taking logs of both sides,

$$(2T-50) \log_{10} 1.08 = \log_{10}(4 \times 10^7)$$

$$2T - 50 = 7.6/0.033 = 227$$

$$T = 139 \text{ °C}$$

For the p-type silicon, $p_0 = 5 \times 10^{21}$ m^{-3} and so $n_0 = 5 \times 10^{20}$ m^{-3}

$$(1.08)^{(2T-50)} = 5 \times 10^{20} \times 5 \times 10^{21} / 2.25 \times 10^{32} = 11.1 \times 10^9$$

$$2T - 50 = \log_{10}(11.1 \times 10^9) / \log 1.08 = 300$$

$$T = 175\,°\text{C}$$

Question 9.4

(a)

The equilibrium density of electrons in the p-region at 25 °C is given by

$$n_{p0} = \frac{n_i^2}{N_a} = \frac{(1.5 \times 10^{16})^2}{3 \times 10^{21}} \text{ m}^{-3} = \frac{2.25 \times 10^{32}}{3 \times 10^{21}} \text{ m}^{-3}$$

$$= 0.75 \times 10^{11} \text{ m}^{-3}$$

This density is increased by the forward bias by the factor

$$\exp(K V_D) = \exp(40 \times 0.5) = e^{20} = 4.85 \times 10^8$$

The electron density next to the transition region is therefore

$$n_p = 0.75 \times 10^{11} \times 4.85 \times 10^8 \text{ m}^{-3} = 3.64 \times 10^{19} \text{ m}^{-3}$$

(b)

If the temperature increases by 20 °C, then n_i^2 and n_{p0} increase by the factor $(1.08)^{40} = 21.7$.

The electron density in the p-region next to the transition region therefore becomes

$$n_p = 3.64 \times 10^{19} \times 21.7 \text{ m}^{-3} = 7.9 \times 10^{20} \text{ m}^{-3}$$

(c)

As in part (a), $n_{p0} = 0.75 \times 10^{11}$ m^{-3}. This is changed by the factor

$$\exp(K V_D) = \exp(40 \times -0.15) = e^{-6} = 2.5 \times 10^{-3}$$

The electron density is therefore

$$n_p = 0.75 \times 10^{11} \times 2.5 \times 10^{-3} \text{ m}^{-3} = 1.9 \times 10^8 \text{ m}^{-3}$$

(d)

With a 20 °C temperature increase, the electron density will be

$$n_p = 1.9 \times 10^8 \times 21.7 \text{ m}^{-3} = 4.1 \times 10^9 \text{ m}^{-3}$$

Question 9.5

(a)

(i) Using the equation $I_D = I_S(e^{KV_D} - 1)$ with $I_S = 3 \times 10^{-13}$ A, $K = 40$ V^{-1} and $V_D = 0.5$ V,

$$I_D = 3 \times 10^{-13} \times (e^{20} - 1) \text{ A} = 3 \times 10^{-13} \times e^{20} \text{ A} \approx 0.15 \text{ mA}$$

(ii) When $V_D = 0.6$ V, $I_D = 3 \times 10^{-13} \times e^{24}$ A ≈ 7.9 mA

(iii) When $V_D = 0.7$ V, $I_D = 3 \times 10^{-13} \times e^{28}$ A ≈ 0.43 A

(iv) When $V_D = 0.8$ V, $I_D = 3 \times 10^{-13} \times e^{32}$ A ≈ 24 A

This last result, while correctly calculated from the formula, must not be taken too seriously. Firstly, with such a voltage and current, the power dissipation in the device would be almost 20 watts, enough to melt all but high power silicon rectifiers. Secondly, the bulk resistance of the silicon from which the device is made would, at such a current, cause voltage drops which would reduce the forward bias of the junction very significantly. Thirdly, for a diode capable of carrying a current higher than tens of milliamps, a value of K of 35 V^{-1} is a more typical value, giving more reasonable values of current as you can see in part (b) of this question.

(b)

With $I_S = 3 \times 10^{-13}$ A and $K = 35$ V^{-1},

(i) When $V_D = 0.5$ V, $I_D = 3 \times 10^{-13} \times e^{17.5}$ A ≈ 12 µA

(ii) When $V_D = 0.6$ V, $I_D = 3 \times 10^{-13} \times e^{21}$ A ≈ 0.4 mA

(iii) When $V_D = 0.7$ V, $I_D = 3 \times 10^{-13} \times e^{24.5}$ A ≈ 13 mA

(iv) When $V_D = 0.8$ V, $I_D = 3 \times 10^{-13} \times e^{28}$ A ≈ 0.43 A

(c)

Using the equation $I_D = I_S(e^{KV_D} - 1)$ with a very large negative value of V_D gives $I_D = -I_S$, so the reverse bias current reaches a maximum value (provided breakdown of the junction does not occur) of $-I_S$.

Hence, to find the diode voltage at which the reverse current reaches 99% of the saturation value, for the diode with $K = 40$ V^{-1},

$$-0.99 I_S = I_S(e^{KV_D} - 1) = I_S(e^{40V_D} - 1)$$
$$e^{40V_D} - 1 = -0.99$$
$$e^{40V_D} = 0.01$$
$$40 V_D = \ln 0.01 \approx -4.6$$
$$V_D \approx -0.12 \text{ V}$$

For the diode with $K = 35$ V^{-1},
$$35V_D = \ln 0.01 \approx -4.6$$
$$V_D = -0.13 \text{ V}$$

Question 9.6

(a)

(i) At 25 °C, $\quad K = \dfrac{q}{kT} = \dfrac{1.6 \times 10^{-19}}{1.38 \times 10^{-23} \times (273+25)}$ V^{-1} = 38.9 V^{-1}

(ii) At 50 °C, $\quad K = \dfrac{1.6 \times 10^{-19}}{1.38 \times 10^{-23} \times (273+50)}$ V^{-1} = 35.9 V^{-1}

(b)

(i) At 25 °C and 0.6 V bias,
$$I_D = I_S(e^{KV_D} - 1) = 3 \times 10^{-13}(e^{38.9 \times 0.6} - 1) \text{ A}$$
$$= 3 \times 10^{-13}(e^{23.34} - 1) \text{ A}$$
$$= 3 \times 10^{-13} \times 1.37 \times 10^{10} \text{ A} = 4.11 \text{ mA}$$

(ii) The saturation current I_S is proportional to $\left(\dfrac{n_{p0}}{l_p} + \dfrac{p_{n0}}{l_n}\right)$ (equation 9.7 on page 382 of the textbook) where n_{p0} and p_{n0} are both proportional to n_i^2.

Hence, for a device where l_p and l_n are unchanging, I_S is proportional to n_i^2.

So, from 25 °C to 50 °C, I_S will increase by a factor of $[(1.08)^{25}]^2 = 46.9$.

Therefore, at 50 °C, $\quad I_S = 3 \times 10^{-13} \times 46.9$ A $= 1.41 \times 10^{-11}$ A
and
$$I_D = 1.41 \times 10^{-11}(e^{35.9 \times 0.6} - 1) \text{ A}$$
$$= 1.41 \times 10^{-11} \times 2.26 \times 10^9 \text{ A} = 31.9 \text{ mA}$$

Question 9.7

(a)

Using equation 9.12 of the textbook (page 392), at $V_{CE} = 10$ V,
$$I_E = I_{SE} e^{KV_{BE}} = 2 \times 10^{-13} \times e^{40 \times 0.65} \text{ A}$$
$$= 2 \times 10^{-13} \times 1.96 \times 10^{11} = 39.1 \text{ mA}$$

Diodes and transistors

Since $\beta = 300$, $\alpha = \beta/(1-\beta) = 0.997$. So, $I_C = \alpha I_E = 39.0$ mA and $I_B = I_C/\beta = 0.13$ mA.

These are the values at $V_{CE} = 10$ V. To find the currents at $V_{CE} = 5$ V we use equation 9.13 of the textbook (page 392) as follows:

$$I_C = \text{constant} \times \frac{V_A + V_{CE}}{V_A}, \text{ and so the constant} = \frac{I_C V_A}{V_A + V_{CE}}.$$

Hence,

$$\frac{I_C \times 120 \text{ V}}{120 \text{ V} + 5 \text{ V}} = \frac{39.0 \text{ mA} \times 120 \text{ V}}{120 \text{ V} + 10 \text{ V}}$$

$$I_C = 39.0 \text{ mA} \times 125/130 = 37.5 \text{ mA}.$$

$I_B = I_C/\beta = 0.125$ mA and $I_E = I_C + I_B = 37.6$ mA.

(b)

The output conductance can be calculated using the equation

$$g_{out} = \frac{I_C}{V_A + V_{CE}} = \frac{37.5 \text{ mA}}{125 \text{ V}} = 0.3 \text{ mS}$$

Hence, the output resistance $r_{out} = 1/g_{out} = (1000/0.3) \, \Omega = 3.3$ kΩ

10 ANALOGUE TRANSISTOR CIRCUITS

QUESTIONS

10.1 (a) A bipolar transistor has $\beta = 400$ and $VA = 60$ V. What are its equivalent circuit parameters when operating at $I_C = 1$ mA and $V_{CE} = 30$ V?

(b) An NPN transistor is specified by the manufacturer as having a β of 100 and an r_o of 11 kΩ when operating at a collector current of 15 mA and a collector-emitter voltage of 10 V. What will be the values of g_m, g_i and g_o when $I_C = 10$ mA and $V_{CE} = 5$ V?

(c) A bipolar PNP transistor has $r_i = 1$ kΩ and $r_o = 35$ kΩ when operating at $I_C = 5$ mA and $V_{CE} = 15$ V. What are the values of β and VA, and what are the values of g_m, g_i and g_o when $I_C = 2$ mA and $V_{CE} = 5$ V?

10.2 A current-mirror current source having the circuit configuration of Figure 10.7(b) of the textbook (page 422) has ± 15 V power supplies and is required to supply a current of 50 µA. The value of R_1 is 20 kΩ; what is the required value of R_E?

10.3 (a) The current-mirror current source of Question 10.2 has R_E changed to 10 kΩ. Show that the value of the output current is defined by the equation

$$\ln\left(I_1 / I_{out}\right) = 4 \times 10^5 I_{out}$$

where I_1 and I_{out} are in amperes, and the value of K for the base-emitter junctions of the transistors is assumed to be 40 V^{-1}.

(b) The value of R_1 is changed to 33 kΩ so that I_1 becomes 890 µA. By completing Table 10.1, and then adding further entries for appropriate values of I_{out}, calculate the value of I_{out} to an accuracy of 1% or better.

I_{out}	I_1/I_{out}	$\ln(I_1/I_{out})$	$4 \times 10^5 I_{out}$
1 µA	890		
10 µA	89		
100 µA	8.9		

Table 10.1

(c) If R_1 is now changed to 10 kΩ and R_E to 15 kΩ, calculate the new value of the output current to an accuracy of 1% or better.

10.4 A common-emitter amplifier having the circuit of Figure 10.11 of the textbook (page 429) has the following component values: $R_E = 1$ kΩ, $R_C = 4.7$ kΩ, $R_1 = 100$ kΩ, $R_2 = 22$ kΩ and

$V_{cc} = 9$ V. The transistor has the following parameters: $\beta = 200$, $g_i = 200$ µS, $g_o = 19$ µS and $K = 40$ V^{-1}.

(a) Calculate the transistor d.c. voltages and currents at the operating point.

(b) Calculate the gain of the amplifier.

(c) If the capacitor at the amplifier input is 1 µF and the capacitor at the emitter is 33 µF, what is the low-frequency 3 dB point of the amplifier? (This is not dealt with in Chapter 10, but your knowledge of Chapter 3 should enable you to perform the relevant calculations.)

10.5 Figure 10.1 shows two versions of a common-emitter amplifer using the same transistor, each of which is required to have a signal voltage gain V_{out}/V_s of at least 100. The transistor has $\beta = 63$, $VA = 100$ and $I_{SE} = 1.5 \times 10^{-14}$ A.

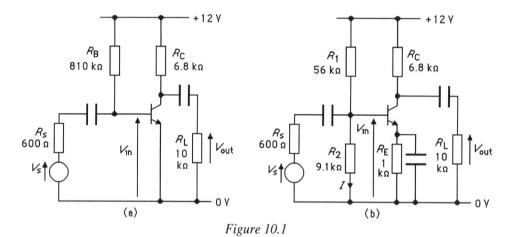

Figure 10.1

(a) Assuming that the operating collector voltage is supposed to be +6 V, perform relevant calculations to check whether either of the circuits satisfies the operating point and voltage gain criteria. The reactances of the capacitors can be considered negligible at all frequencies of interest.

(b) Calculate the minimum value of the input capacitor for each circuit if the low frequency 3 dB point is to be less than 200 Hz. (Assume that the reactances of the output capacitor and the emitter capacitor are still negligible at 200 Hz.)

10.6 A signal whose source has negligible resistance is to be amplified using the circuit of Figure 10.9(b) of the textbook (p. 425) with $R_L = 1$ kΩ, $R_C = 1.2$ kΩ. and $V_{CC} = 10$ V. The transistor has the following parameters: $\beta = 300$, $g_i = 560$ µS, $g_o = 25$ µS and $K = 40$ V^{-1}.

(a) Choose a suitable value for the resistor R_B and calculate the gain of the amplifier. What is the maximum output signal amplitude before clipping occurs?

(b) What new gain would result from increasing R_C to 1.8 kΩ without changing any other circuit components? What would be the new maximum output signal amplitude before clipping?

(c) Calculate the new value of R_B required to set the d.c. collector voltage to 5 V with this increased value of R_C. Re-calculate the gain of the amplifier. Comment on the result obtained.

10.7 (a) If, in the circuit of Question 10.6, the circuit configuration and the value of the load resistor are fixed, and the transistor cannot be changed, what type of changes to V_{cc}, R_C and R_B will allow the gain to be increased to at least 120 when the required maximum output signal amplitude (without clipping) is 4.8 V? Calculate suitable new values for the changed components to give this new gain.

(b) Instead of the component changes suggested in (a), the gain is to be increased by using a dynamic load on the transistor as in the circuit of Figure 10.2. What will be the gain of this circuit? (Assume that T1 and T3 have the same output conductance g_o.)

(c) What simple addition to the circuit of Figure 10.2 will change the gain to 120?

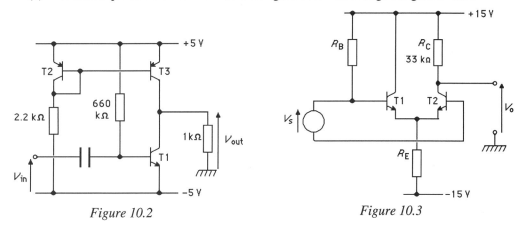

Figure 10.2   Figure 10.3

10.8 (a) A long-tailed pair amplifier has the configuration shown in Figure 10.3. The output voltage V_o is required to be 0 V when V_s is zero, and to have at least a ±5 V swing when the input signal is present. The transistors have $VA = 110$ V and $\beta = 60$. The signal source has negligible internal resistance.

(i) Suggest a suitable value for the operating-point base voltage of the transistors.

(ii) Calculate the value of R_E to give an operating point output voltage of 0 V.

(iii) Calculate the required value of R_B.

(iv) Calculate the gain of the amplifier.

(b) The tail resistor in the long-tailed pair is now replaced by the current source shown in Figure 10.4.

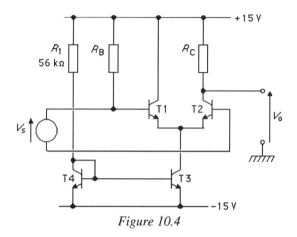

Figure 10.4

(i) If the operating point output voltage is to remain 0 V, calculate the new values of R_C and R_B.

(ii) Calculate the new gain of the amplifier.

(c) A load resistor of 5 kΩ is now connected between the amplifier output and the 0 V line.

(i) Calculate the new amplifier gain.

(ii) For signals, as far as the load is concerned, the circuit of Figure 10.4 can be represented by a Thévenin or Norton equivalent circuit. Evaluate the Thévenin and Norton equivalent circuit components in terms of V_s.

10.9 (a) The transistor in the circuit of Figure 10.5 has $\beta = 100$ and $VA = 90$ V. What value of R_B will give an output d.c. voltage of 0 V?

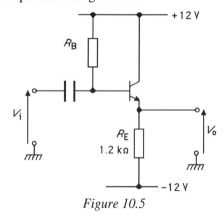

Figure 10.5

(b) If $R_B = 120$ kΩ, calculate the output d.c. voltage.

(c) The transistor type is specified by the manufacturer as having a typical value of β of 100, but with different devices having a range of values of β from 50 to 200. If R_B is to made up of the series connection of a resistor and a pot (connected as a variable resistor) so as to be able to adjust the output d.c. voltage to zero whatever the β of the transistor (within the specified range), what is the minimum value of the pot's resistance?

(d) If R_B is fixed at 120 kΩ, what is the range of values of the output voltage which could be obtained with this transistor type?

(e) Calculate the input and output resistance of the circuit when R_B is 115 kΩ and $\beta = 100$, given that a load of 500 Ω is attached, that the resistance of the source of V_i is 10 kΩ and that the reactance of the capacitor is negligible.

10.10 In the amplifier circuit of Figure 10.6, $I_1 = 200$ µA, $I_2 = 2$ mA and $I_3 = 20$ mA. The output voltage is zero when the input voltage is zero. The transistors have the following parameters:

NPN transistors, $\beta = 100$, $VA = 100$ V
PNP transistors, $\beta = 200$, $VA = 60$ V

(a) Calculate the values of all resistors in the circuit.

(b) Calculate the voltage gain of each stage of the amplifier and hence calculate the overall voltage gain of the amplifier when the source resistance of the input signal is negligible.

(c) Calculate the input and output resistance of the amplifier.

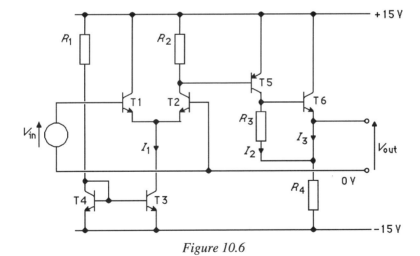

Figure 10.6

Analogue transistor circuits

SOLUTIONS

Question 10.1

(a)

$$g_m = KI_C = 40 \text{ V}^{-1} \times 10^{-3} \text{ A} = 40 \text{ mA V}^{-1}$$

$$g_i = g_m/\beta = (40/400) \text{ mS} = 100 \text{ }\mu\text{S}$$

$$g_o = \frac{I_C}{V_A + V_{CE}} = \frac{10^{-3} \text{ A}}{(60+30) \text{ V}} = 11 \text{ }\mu\text{S}$$

(b)

From the data given, since $r_o = 1/g_o = \dfrac{V_A + V_{CE}}{I_C}$, $V_A = I_C r_o - V_{CE}$. Hence,

$$V_A = 15 \times 10^{-3} \text{ A} \times 11 \times 10^3 \text{ }\Omega - 10 \text{ V} = 155 \text{ V}$$

At $I_C = 10$ mA, $V_{CE} = 5$ V,

$$g_m = KI_C = 40 \text{ V}^{-1} \times 10^{-2} \text{ A} = 400 \text{ mA V}^{-1}$$

$$g_i = g_m/\beta = (400/100) \text{ mS} = 4 \text{ mS}$$

$$g_o = \frac{I_C}{V_A + V_{CE}} = \frac{10^{-2} \text{ A}}{(155+5) \text{ V}} = 63 \text{ }\mu\text{S}$$

(c)

At $I_C = 5$ mA, $V_{CE} = 15$ V,

$$g_m = KI_C = 40 \text{ V}^{-1} \times 5 \times 10^{-3} \text{ A} = 200 \text{ mA V}^{-1}$$

$$g_i = g_m/\beta \text{ so } \beta = g_m/g_i = g_m r_i = 200 \text{ mA V}^{-1} \times 10^3 \text{ }\Omega = 200$$

$$V_A = I_C r_o - V_{CE} = 5 \times 10^{-3} \text{ A} \times 35 \times 10^3 \text{ }\Omega - 15 \text{ V} = 160 \text{ V}$$

At $I_C = 2$ mA, $V_{CE} = 5$ V,

$$g_m = KI_C = 40 \text{ V}^{-1} \times 2 \times 10^{-3} \text{ A} = 80 \text{ mA V}^{-1}$$

$$g_i = g_m/\beta = (80/200) \text{ mS} = 400 \text{ }\mu\text{S} \text{ or } r_i = 2.5 \text{ k}\Omega$$

$$g_o = \frac{I_C}{V_A + V_{CE}} = \frac{2 \times 10^{-3} \text{ A}}{(160+5) \text{ V}} = 12 \text{ }\mu\text{S} \text{ or } r_o = 82.5 \text{ k}\Omega$$

Question 10.2

The voltage drop across $R_1 = 30$ V $- 0.65$ V $= 29.35$ V.

With $R_1 = 20$ kΩ, $I_1 = 29.35$ V $\div 20$ k$\Omega = 1.47$ mA.

When the output current is 50 μA, $I_1/I_{out} = 1470/50 = 29.4$.

Using the equation $I_C \approx I_E = I_{SE}(e^{KV_{BE}} - 1) \approx I_{SE}e^{KV_{BE}}$ for a forward biassed base-emitter junction,

$$I_1 = I_{SE}e^{KV_{BE(1)}} \text{ and } I_{out} = I_{SE}e^{K[V_{BE(1)} - V_{R_E}]} = I_{SE}\left(e^{KV_{BE(1)}} / e^{KV_{R_E}}\right)$$

So,

$$I_1/I_{out} = e^{KV_{R_E}}$$

$$KV_{R_E} = \ln(I_1/I_{out}) = \ln 29.4 = 3.38$$

$$V_{R_E} = 3.38/(40\text{ V}^{-1}) = 0.085 \text{ V}$$

Hence,

$$R_E = 0.085 \text{ V} \div 50 \text{ μA} = 1.69 \text{ kΩ}$$

Question 10.3

(a)

Using the equation $I_C \approx I_{SE}e^{KV_{BE}}$ for a forward biassed base-emitter junction, the solution to Question 10.2 has shown that $\ln(I_1/I_{out}) = KV_{R_E}$ where $V_{R_E} = I_{out} \times R_E = 10^4 \times I_{out}$.

Hence, if $K = 40$ V^{-1}, $\quad \ln(I_1/I_{out}) = 4 \times 10^5 I_{out}$ which is the required result.

(b)

The completed table is shown below

I_{out}	I_1/I_{out}	$\ln(I_1/I_{out})$	$4 \times 10^5 I_{out}$
1 μA	890	6.79	0.4
10 μA	89	4.49	4.0
100 μA	8.9	2.19	40

Clearly, the value of I_{out} at which the entries in the last two columns would become equal will lie between 10 μA and 100 μA, and very close to 10 μA. An initial estimate would be 11 μA, since that value will make the entry in the last column equal to 4.4 and will slightly reduce the entry in the last-but-one column. This gives:

I_{out}	I_1/I_{out}	$\ln(I_1/I_{out})$	$4 \times 10^5 I_{out}$
1 μA	890	6.79	0.4
10 μA	89	4.49	4.0
100 μA	8.9	2.19	40
11 μA	80.9	4.39	4.40

Analogue transistor circuits

This looks close enough to satisfy the 1% criterion, but to be sure, you should calculate the entries for 11.1 μA and 10.9 μA. The completed table is shown below and clearly establishes that the output current will be 11.0 μA to an accuracy of better than 1%.

I_{out}	I_1/I_{out}	$\ln(I_1/I_{out})$	$4 \times 10^5 \, I_{out}$
1 μA	890	6.79	0.4
10 μA	89	4.49	4.0
100 μA	8.9	2.19	40
11 μA	80.9	4.39	4.40
11.1 μA	80.2	4.384	4.44
10.9 μA	81.7	4.402	4.36

(c)

With $R_1 = 10$ kΩ, $I_1 = 29.35$ V ÷ 10 kΩ = 2.935 mA.

With $R_E = 15$ kΩ, the equation for I_{out} is

$$\ln(I_1/I_{out}) = KV_{R_E} = 40 \text{ V}^{-1} \times 1.5 \times 10^4 \, \Omega \times I_{out} = 6 \times 10^5 \, I_{out}$$

Completing a table similar to that of part (b) gives, initially

I_{out}	I_1/I_{out}	$\ln(I_1/I_{out})$	$6 \times 10^5 \, I_{out}$
1 μA	2935	7.98	0.6
10 μA	293.5	5.68	6.0
100 μA	29.35	3.38	60

Clearly, the output current will be just less than 10 μA. so we first try 9 μA. The result is shown in the fourth line of the table below. The actual value is somewhere between 9 and 10 μA and so we try 9.5 μA. The result is shown in the fifth line of the table. The last two column entries agree to within 1%, but to confirm the required accuracy we try values of output current about 1% either side of this value. The results from these values suggest that the actual value of the output current is about 9.55 μA, so 9.5 μA is within 1% of the actual value.

I_{out}	I_1/I_{out}	$\ln(I_1/I_{out})$	$6 \times 10^5 \, I_{out}$
1 μA	2935	7.98	0.6
10 μA	293.5	5.68	6.0
100 μA	29.35	3.38	60
9 μA	326.1	5.79	5.40
9.5 μA	308.9	5.73	5.70
9.6 μA	305.7	5.72	5.76
9.4 μA	312.2	5.74	5.64

Question 10.4

(a)

As a first approximation, assume that the current through R_1 and R_2 is much greater than the base current of the transistor (we will check this assumption later).

The current through R_1 and R_2 is therefore $I = 9\text{ V}/(100 + 22)\text{ k}\Omega = 74\text{ }\mu\text{A}$.

The voltage at the base of the transistor is therefore $9\text{ V} \times 22\text{ k}\Omega \div 122\text{ k}\Omega = 1.62\text{ V}$.

Since the base-emitter voltage of the transistor will be very close to 0.65 V, the emitter voltage must be close to 0.97 V. The current through R_E must therefore be $0.97\text{ V} \div 1\text{ k}\Omega = 0.97\text{ mA}$. Since the current gain β is large (200), this is also the value of the collector current.

The collector voltage is therefore $(9\text{ V} - 0.97\text{ mA} \times 4.7\text{ k}\Omega) = 4.4\text{ V}$.

The base current of the transistor will be $0.97\text{ mA} \div 200 \approx 5\text{ }\mu\text{A}$, which is much less than the 74 μA flowing through R_1 and R_2.

[If you wish, the values of the voltages and currents at the operating point can be re-calculated, allowing for the base current. The calculation proceeds much as the one above, except that the calculation of the base voltage of the transistor is slightly more complicated.]

If the current through R_2 is assumed to be I μA then the current through R_1 is $(I + 5)$ μA and the equation for the bias resistor chain becomes:

$$(I+5) \times 10^{-6}\text{ A} \times 10^5\text{ }\Omega + I \times 10^{-6}\text{ A} \times 0.22 \times 10^5\text{ }\Omega = 9\text{ V}$$

This gives $I = 70\text{ }\mu\text{A}$ and hence a base voltage of $70\text{ }\mu\text{A} \times 22\text{ k}\Omega = 1.54\text{ V}$.

Repeating the calculation as above gives: $V_B = 0.89\text{ V}$, $I_E = I_C = 0.89\text{ mA}$, $V_C = 4.8\text{ V}$.]

(b)

Using the equivalent circuit of Figure 10.10 of the textbook (page 426), the current equation at node 2, neglecting g_L is:

$$g_m V_{in} + V_{out} g_o + V_{out} g_C = 0$$

and so the gain of the amplifier is given by:

$$A_V = V_{out}/V_{in} = -\frac{g_m}{g_o + g_C}$$

Now $g_m \approx K I_C$, where $K \approx 40$ V^{-1}, and $I_C = 0.97$ mA (using the approximate values calculated first in part (a)), so

$$A_V = V_{out}/V_{in} = -\frac{40\text{ V}^{-1} \times 0.97\text{ mA}}{19\text{ }\mu\text{S} + 213\text{ }\mu\text{S}} = -167$$

Analogue transistor circuits

(c)

The input capacitor, together with the conductances g_1, g_2 and g_i in parallel, form a first-order, high-pass, CR filter.

The filter time constant is $\tau = \dfrac{C}{g_1 + g_2 + g_i} = \dfrac{10^{-6}\text{ F}}{(10 + 45 + 200)\,\mu\text{S}} = 3.9\text{ ms}$

The 3 dB frequency for the filter is the frequency at which $\omega_c \tau = 1$, i.e at the frequency:

$$f_c = \omega_c/2\pi = 1/2\pi\tau = 1 \Big/ \left(2\pi \times 3.9 \times 10^{-3}\text{ s}\right) = 41\text{ Hz}$$

The emitter capacitor will also cause the amplifier gain to fall off at low frequencies because, as its reactance becomes comparable with the resistance of R_E, the amount of negative feedback for signals will increase, so reducing the gain.

The frequency at which the reactance of the emitter capacitor becomes equal to the resistance R_E is given by:

$$1/2\pi fC = R_E \text{ so that } f = 1/2\pi C R_E = \dfrac{1}{2\pi \times 33 \times 10^{-6}\text{ F} \times 10^3\,\Omega} \approx 4.8\text{ Hz}$$

This is almost a decade below the 3 dB frequency of the input circuit, so the effect of the emitter capacitor does not contribute significantly to the amplifier low-frequency 3 dB point, which is therefore 41 Hz.

Question 10.5

(a)

For circuit (a), the base current will be $(12 - 0.65)\text{ V} \div 810\text{ k}\Omega = 14\,\mu\text{A}$.

With $\beta = 63$, the collector current will be $14\,\mu\text{A} \times 63 = 0.88\text{ mA}$.

The collector voltage will be $12\text{ V} - 0.88\text{ mA} \times 6.8\text{ k}\Omega = 6.0\text{ V}$.

The d.c. operating point of the transistor is correct.

To find the gain of the amplifier, we need the values of g_o, g_i and g_m.

$$g_o = \dfrac{I_C}{V_A + V_{CE}} = \dfrac{0.88 \times 10^{-3}\text{ mA}}{(100 + 6)\text{ V}} = 8.3\,\mu\text{S}$$

$$g_m = KI_C = 40\text{ V}^{-1} \times 0.88 \times 10^{-3}\text{ A} = 35\text{ mA V}^{-1}$$

$$g_i = g_m/\beta = (35/63)\text{ mS} = 560\,\mu\text{S}$$

The gain $\dfrac{V_{out}}{V_{in}} = -\dfrac{g_m}{g_o + g_C + g_L} = -\dfrac{35\times 10^{-3}\text{ A V}^{-1}}{(8.3+147+100)\,\mu\text{S}} = 137$

The amplifier gain $\dfrac{V_{out}}{V_s} = \dfrac{V_{out}}{V_{in}} \times \dfrac{V_{in}}{V_s}$ so we need to find V_{in}/V_s, and to find this ratio we need to know the input resistance of the amplifer. This input resistance will be the reciprocal of the sum of g_i and g_B, i.e.

Input resistance = $R_{in} = 1/(560+1.2)\,\mu\text{S} = 1.8\text{ k}\Omega$

Hence, $\dfrac{V_{in}}{V_s} = \dfrac{1.8\text{ k}\Omega}{1.8\text{ k}\Omega + 600\,\Omega} = 0.75$ and $\dfrac{V_{out}}{V_s} = \dfrac{V_{out}}{V_{in}} \times \dfrac{V_{in}}{V_s} = 137 \times 0.75 = 103$

Hence the amplifier configuration of circuit (a) has sufficient gain to satisfy the specification.

For circuit (b), assume, as in Question 10.4, that the current through the base bias resistor chain is much greater than the base current. This current will then be:

$$I = 12\text{ V}/(56+9.1)\text{ k}\Omega = 184\,\mu\text{A}.$$

The voltage at the base of the transistor is therefore $12\text{ V} \times 9.1\text{ k}\Omega \div 65.1\text{ k}\Omega = 1.68\text{ V}$.

Since the base-emitter voltage of the transistor will be very close to 0.65 V, the emitter voltage must be close to 1.03 V. The current through R_E must therefore be $1.03\text{ V} \div 1\text{ k}\Omega = 1.03\text{ mA}$. Since the current gain β is 63, $\alpha = \beta/(1+\beta) = 0.98$ and the collector current $I_C = 1.03\text{ mA} \times 0.98 = 1\text{ mA}$.

We can now re-calculate, allowing for the base current of the transistor.

The base current of the transistor will be $1.0\text{ mA} \div 63 \approx 16\,\mu\text{A}$.

The equation for the current I through resistor R_2 of the base bias voltage chain (I in mA) will now be:

$$56\text{ k}\Omega \times (I + 0.016)\text{ mA} + 9.1\text{ k}\Omega \times I\text{ mA} = 12\text{ V}$$

$$65.1\text{ k}\Omega \times I = 12\text{ V} - 56\text{ k}\Omega \times 0.016\text{ mA} = 11.1\text{ V}$$

$$I = 11.1\text{ V} \div 65.1\text{ k}\Omega = 170\,\mu\text{A}$$

The base voltage is then $170\,\mu A \times 9.1\text{ k}\Omega = 1.55\text{ V}$, and the emitter voltage is $1.55\text{ V} - 0.65\text{ V} = 0.90\text{ V}$.

This gives a more accurate value for I_E of 0.9 mA and for I_C of $0.9\text{ mA} \times 0.98 = 0.88\text{ mA}$.

The collector voltage will then be $0.88\text{ mA} \times 6.8\text{ k}\Omega = 6\text{ V}$ as required by the specification.

The values of g_m, g_i and g_o will be the same as for circuit (a), as will g_C and g_L so that the gain V_{out}/V_{in} will be 137 as before.

The input resistance of the amplifier will this time be the reciprocal of the sum of g_i, g_1 and g_2, where $g_i = 560\,\mu\text{S}$, $g_1 = 1/R_1 = 18\text{ mS}$ and $g_2 = 1/R_2 = 110\,\mu\text{S}$.

Analogue transistor circuits

i.e. Input resistance = $R_{in} = 1/(560 + 18 + 110)\,\mu S = 1.45\,k\Omega$

Hence, $\dfrac{V_{in}}{V_s} = \dfrac{1.45\,k\Omega}{1.45\,k\Omega + 600\,\Omega} = 0.71$ and $\dfrac{V_{out}}{V_s} = \dfrac{V_{out}}{V_{in}} \times \dfrac{V_{in}}{V_s} = 137 \times 0.71 = 97$

Hence the amplifier configuration of circuit (b) does not quite have sufficient gain to satisfy the specification, but is very close to it.

(b)

For the circuit of Figure 10.1(a), the input circuit time constant is

$$\tau = C \times \left(\dfrac{1}{g_i + g_B} + R_s\right) = C \times \left(\dfrac{1}{561\,\mu S} + 600\,\Omega\right) = C \times 2.38\,k\Omega$$

If the low-frequency 3 dB point is to be 200 Hz = 400π rad s^{-1}, then $400\pi\tau = 1$.

$$400\pi \times C \times 2.38\,k\Omega = 1$$

$$C = 1/(400\pi \times 2.38\,k\Omega) = 0.33\,\mu F$$

So C must be at least 0.33 µF.

For the circuit of Figure 10.1(b), the input circuit time constant is

$$\tau = C \times \left(\dfrac{1}{g_i + g_1 + g_2} + R_s\right) = C \times \left(\dfrac{1}{(560 + 18 + 110)\,\mu S} + 600\,\Omega\right)$$

$$= C \times (1.45\,k\Omega + 600\,\Omega)$$

$$= C \times 2.05\,k\Omega$$

Hence $C \geq 1/(400\pi \times 2.05\,k\Omega) = 0.39\,\mu F$

Question 10.6

(a)

Unless otherwise specified, the operating d.c. collector voltage of a transistor amplifier is approximately half the supply voltage to allow equal signal amplitude either side of the operating point. In this case therefore, the required d.c. collector voltage is +5 V.

The operating collector current is then $I_C = 5\,V \div 1.2\,k\Omega = 4.2\,mA$.

The required base current is therefore 4.2 mA ÷ 300 = 14 µA.

So, required value of $R_B = (10\,V - 0.65\,V) \div 14\,\mu A = 670\,k\Omega$.

Using the equivalent circuit of Figure 10.10(b) of the textbook,

$$\frac{V_{out}}{V_{in}} = \frac{V_{out}}{V_s} = -\frac{g_m}{g_o + g_C + g_L} = -\frac{KI_C}{g_o + g_C + g_L}$$

Hence,
$$\frac{V_{out}}{V_S} = -\frac{40 \text{ V}^{-1} \times 4.2 \times 10^{-3} \text{ A}}{(25 + 833 + 1000) \text{ μS}} = -90$$

Clipping of the output waveform will occur if the signal attempts to drive the collector voltage above the supply voltage V_{CC} or below the minimum collector voltage (which is about 0.2 V as described in Chapter 11 of the textbook).

When the operating point collecter voltage is half the supply voltage, the maximum collector signal voltage is the same for both positive and negative excursions, and this is therefore the normal optimum operating point. With $V_{CC} = 10$ V and the d.c. collector voltage = 5 V, the maximum output signal amplitude without clipping is about 5 V.

(b)

If the value of R_C is increased to 1.8 kΩ with no change in other circuit components, the gain will increase to

$$\frac{V_{out}}{V_S} = -\frac{40 \text{ V}^{-1} \times 4.2 \times 10^{-3} \text{ A}}{(25 + 556 + 1000) \text{ μS}} = -106$$

However, the d.c. collector voltage will have changed to (10 V − 1.8 kΩ × 4.2 mA) = 2.44 V so that clipping of the output waveform will occur at an amplitude of about 2.4 V.

(c)

To set the d.c. collector voltage to 5 V, the collector current must change to 5 V ÷ 1.8 kΩ = 2.8 mA. This means that the base current must change to 2.8 mA ÷ 300 ≈ 9.3 μA which requires R_B to increase to 9.35 V ÷ 9.3 μA = 1 MΩ

At the new collector current of 2.8 mA, the voltage gain of the amplifier is given by:

$$\frac{V_{out}}{V_S} = -\frac{40 \text{ V}^{-1} \times 2.8 \times 10^{-3} \text{ A}}{(25 + 556 + 1000) \text{ μS}} = -71$$

The increase in the value of R_C (while retaining the same collector operating voltage) has caused a decrease in the overall gain magnitude (from 90 to 71) rather than an expected increase. This is because (i) the collector current has had to be reduced to preserve the operating point and hence the output voltage swing, and this decrease in I_C causes a proportional decrease in g_m, and (ii) the small signal output resistance of the amplifier feeding R_L has increased, so increasing the attenuation when the load is connected (this is accounted for in the maths by the presence of g_L in the equation for the amplifier gain).

Question 10.7

(a)

From the results of Question 10.6 part (b), it seems that an increase in R_C, provided it is not accompanied by a reduction in I_C, will give an increase in the voltage gain of the amplifier.

Analogue transistor circuits

To achieve this, V_{CC} needs to be increased as well, so that the increased d.c. voltage drop across R_C does not reduce the collector voltage below the level necessary to achieve the required output voltage signal amplitude.

For an output signal amplitude of 4.8 V, a d.c. collector voltage of 5 V is a minimum requirement.

For example, if a V_{CC} of 15 V is used, R_C could be increased until the voltage drop across it is 10 V when the original collector current of 4.2 mA is flowing through it. This would give a value for R_C of 10 V ÷ 4.2 mA = 2.4 kΩ and a gain of

$$\frac{V_{out}}{V_S} = -\frac{40 \text{ V}^{-1} \times 4.2 \times 10^{-3} \text{ A}}{(25 + 417 + 1000) \text{ μS}} = -116$$

which is close to the required value.

With the same value of V_{CC}, a lower value of R_C, combined with an increase in the collector d.c. current I_C would give even greater gain. For example, if $R_C = 2$ kΩ, and the voltage drop across R_C is still to be 10 V, the collector current can be increased to 5 mA (by reducing the value of R_B), so giving a gain of:

$$\frac{V_{out}}{V_S} = -\frac{40 \text{ V}^{-1} \times 5 \times 10^{-3} \text{ A}}{(25 + 500 + 1000) \text{ μS}} = -131$$

Clearly, there is a value of R_C which will satisfy the specification, and to determine that value without resorting to trial and error, we can use a mathematical approach. At the same time, I will revert to the original concept of the optimum collector voltage being just half the supply voltage.

For a small-signal gain of exactly 120,

$$\frac{V_{out}}{V_S} = -120 = -\frac{40 \text{ V}^{-1} \times I_C}{(25 \text{ μS} + g_c + 1000 \text{ μS})}$$

which gives

$$I_C = 3 \text{ V} \times (25 \text{ μS} + g_c + 1000 \text{ μS})$$
$$= 3 \text{ V} \times 1025 \text{ μS} + 3 \text{ V} \times g_c$$
$$= 3.075 \text{ mA} + 3 \text{ V} \times g_c$$

Re-arranging, we get $\quad g_c = \dfrac{I_C - 3.075 \text{ mA}}{3 \text{ V}} \quad$ or $\quad R_C = \dfrac{3 \text{ V}}{I_C - 3.075 \text{ mA}} \quad\quad (1)$

If the d.c. collector voltage is to be half the supply voltage, then $\quad I_C R_C = V_{CC}/2$

From equation 1, $\quad 3 \text{ V} = V_{CC}/2 - R_C \times 3.075 \text{ mA}$

which can be re-arranged to give $\quad R_C = \dfrac{V_{CC} - 6 \text{ V}}{6.15 \text{ mA}}$

So, for any chosen value of V_{CC} greater than 6 V, a value of R_C can be found which will give the amplifier the required gain of 120. However, to meet the output signal amplitude criterion, we have already established that the d.c. collector voltage must be at least 5 V, which means that the supply must be at least 10 V, so the equation must only be used for values of V_{CC} of 10 V or more.

In Table 10.2 I have performed the calculation for value of V_{CC} of 10, 12, 15 and 20 V, have also calculated the required new value of R_B and checked the resulting amplifier gain. Rounding of values obtained results in the magnitude of the gain apparently not being exactly 120.

V_{CC}/V	R_C/Ω	I_C/mA	R_B/kΩ	A_V
10	650	7.7	360	−123
12	980	6.2	550	−121
15	1500	5.1	840	−121
20	2300	4.4	1320	−121

Table 10.2

(b)

The current mirror formed by T2 and T3 will generate a constant current into the collector of T3, with a very low small signal output conductance (very high output resistance). The value of the constant current is given by the equation $I = (10\text{ V} - 0.65\text{ V}) \div 2.2\text{ k}\Omega = 4.25\text{ mA}$.

The base resistor of T1 fixes its collector current at the value $I_C = 300 \times (10\text{ V} - 0.65\text{ V}) \div 660\text{ k}\Omega$ = 4.25 mA which is the current supplied by T3. The current through R_L is therefore zero, and the d.c. volt drop across it is zero.

As described on page 433 of the textbook, an input signal will cause a change in the collector current of T1, but not in the current flowing through T3. The difference between the two currents must flow through R_L, so creating the output signal voltage.

As stated on page 434 of the textbook, the equation for the gain is

$$\frac{V_{out}}{V_{in}} = -\frac{g_m}{g_{o1} + g_{o3} + g_L} = -\frac{KI_C}{g_{o1} + g_{o3} + g_L}$$

and if T1 and T3 have the same output conductance then $g_{o3} = g_{o1} = 25$ μS and the expression for the gain becomes:

$$\frac{V_{out}}{V_{in}} = -\frac{40\text{ V}^{-1} \times 4.25 \times 10^{-3}\text{ A}}{(25 + 25 + 1000)\text{ μS}} = -162$$

(c)

To reduce the gain to 120 without affecting the d.c. conditions of the circuit, a resistor R_2 can be placed across R_L. This resistor will shunt the load and so allow some of the signal current, generated as the difference between the collector currents of T1 and T3, to bypass the load. Mathematically, the presence of this resistor introduces another conductance into the denominator of the expression for the gain, and so reduces the gain.

To find the required value of R_2, we use the equation

$$\frac{V_{out}}{V_{in}} = -120 = -\frac{40\text{ V}^{-1} \times 4.25 \times 10^{-3}\text{ A}}{g_2 + (25 + 25 + 1000)\text{ μS}}$$

which gives the result $g_2 = 370$ μS or $R_2 = 2.7$ kΩ.

Analogue transistor circuits

Question 10.8

(a)

(i) If the output voltage is to be able to change by ±5 V from 0 V, since the collector of T2 cannot become more negative than the base, the base operating point voltage must be at most −5 V, and −6 V would be safer to avoid the risk of clipping.

(ii) If the collector voltage at the operating point is 0 V, the operating point collector current must be 15 V ÷ 33 kΩ = 0.45 mA. The total current through the tail resistor is then 0.9 mA.

If the bases are at −6 V, the emitters will be at −6.65 V to make V_{BE} = 0.65 V, so R_E = (15 V − 6.65 V) ÷ 0.9 mA = 9.3 kΩ.

(iii) If I_C = 0.45 mA, then I_B = 0.45 mA ÷ 60 = 7.5 μA. The total current through R_B will therefore be 15 μA and hence R_B = (15 V + 6 V) ÷ 15 μA = 1.4 MΩ.

(iv) The gain = $\dfrac{V_o}{V_s} = \tfrac{1}{2} \times \dfrac{g_m}{g_o + g_C}$

where
$$g_m = KI_C = 40 \text{ V}^{-1} \times 0.45 \text{ mA} = 18 \text{ mA V}^{-1}$$
$$g_o = \dfrac{I_C}{V_A + V_{CE}} = \dfrac{0.45 \text{ mA}}{(110 + 6.65) \text{ V}} = 3.9 \text{ μS}$$
$$g_C = 1/R_C = 1/33 \text{ kΩ} = 30 \text{ μS}$$

So, $\dfrac{V_o}{V_s} = \tfrac{1}{2} \times \dfrac{18 \times 10^{-3}}{34 \times 10^{-6}} = 265$

(b)

(i) The current through R_1 = (30 V − 0.65 V) ÷ 56 kΩ = 0.52 mA. Hence the current in each emitter = 1/2 × 0.52 mA = 0.26 mA.

For the output operating point voltage to be 0 V,
$$R_C = 15 \text{ V} ÷ 0.26 \text{ mA} = 58 \text{ kΩ}.$$

The base current in each transistor = 0.26 mA ÷ 60 = 4.3 μA. The current through R_B is therefore 8.6 μA and
$$R_B = 21 \text{ V} ÷ 8.6 \text{ μA} = 2.4 \text{ MΩ}.$$

(ii) $g_m = KI_C = 40 \text{ V}^{-1} \times 0.26 \text{ mA} = 10.4 \text{ mA V}^{-1}$

$$g_o = \dfrac{I_C}{V_A + V_{CE}} = \dfrac{0.26 \text{ mA}}{(110 + 6.65) \text{ V}} = 2.2 \text{ μS}$$

$$g_C = 1/R_C = 1/58 \text{ kΩ} = 17 \text{ μS}$$

$$\dfrac{V_o}{V_s} = \tfrac{1}{2} \times \dfrac{g_m}{g_o + g_C} = \tfrac{1}{2} \times \dfrac{10.4 \times 10^{-3}}{19 \times 10^{-6}} = 274$$

(c)

(i) If a load resistor $R_L = 5$ kΩ is now connected to the output, the gain is:

$$\frac{V_o}{V_s} = \tfrac{1}{2} \times \frac{g_m}{g_o + g_C + g_L} = \tfrac{1}{2} \times \frac{10.4 \times 10^{-3}}{19 \times 10^{-6} + 200 \times 10^{-6}} \approx 24$$

(ii) For the Thévenin equivalent circuit, the voltage source V_T is the output voltage with the load removed, which is $274 \times V_s$. The resistance R_T is the output resistance of the amplifier without R_L which is $1/(g_o + g_C)$.

Hence, $\quad V_T = 274 V_s$ and $R_T = \dfrac{1}{g_o + g_C} = \dfrac{1}{19 \times 10^{-6}\text{ S}} = 53$ kΩ

For the Norton equivalent circuit,

$I_N = V_T / R_T = 274 V_s / 53$ k$\Omega = 5.2$ mS $\times V_s$
$R_N = R_T = 53$ kΩ

Question 10.9

(a)

If the d.c. output voltage is zero, then $I_E = 12$ V $\div 1.2$ k$\Omega = 10$ mA.

Since $\beta = 100$, $\alpha = 0.99$ and so $I_C = 9.9$ mA and $I_B = 99$ µA.

Hence, $R_B = (12$ V $- 0.65$ V$) \div 99$ µA $= 115$ kΩ.

(b)

If $R_B = 120$ kΩ and the d.c. output voltage is V_o, then, in the emitter circuit,

$$V_o - (-12\text{ V}) = I_E \times 1.2 \text{ k}\Omega$$

$$I_E = \frac{V_o + 12 \text{ V}}{1.2 \text{ k}\Omega}$$

In the base circuit, $\quad 12$ V $- (V_o + 0.65$ V$) = I_B \times 120$ kΩ

But $\quad I_B = I_C/\beta \approx I_E/\beta = \dfrac{V_o + 12 \text{ V}}{1.2 \text{ k}\Omega \times 100}$

So,

$$12 \text{ V} - (V_o + 0.65 \text{ V}) = I_B \times 120 \text{ k}\Omega = V_o + 12 \text{ V}$$

$$-0.65 \text{ V} = 2 V_o$$

$$V_o = -0.325 \text{ V}.$$

Analogue transistor circuits

(c)

When the output d.c. voltage is zero and $\beta = 50$, $R_B = 11.35 \text{ V} \div I_B$, where $I_B \approx I_E/\beta = 20 \text{ μA}$.

Hence, $R_B = 11.35 \text{ V} \div 20 \text{ μA} = 0.57 \text{ M}\Omega$.

When the output d.c. voltage is zero and $\beta = 200$, $I_B \approx I_E/\beta = 5 \text{ μA}$ and $R_B = 11.35 \text{ V} \div 5 \text{ μA} = 2.3 \text{ M}\Omega$.

The base bias resistor must therefore consist of a fixed 560 kΩ resistor, together with a 1.8 MΩ (or more probably 2 MΩ) pot.

(d)

When $R_B = 120$ kΩ and $\beta = 50$, the output voltage can be calculated from the equation:

$$12 \text{ V} - (V_o + 0.65 \text{ V}) = I_B \times 120 \text{ k}\Omega = \frac{(V_o + 12 \text{ V})}{50 \times 1.2 \text{ k}\Omega} \times 120 \text{ k}\Omega$$

$$= (V_o + 12 \text{ V}) \times 2$$

Hence, $(12 - 24 - 0.65) \text{ V} = 3V_o$ and $V_o = -12.65 \text{ V}/3 = -4.22 \text{ V}$

When $R_B = 120$ kΩ and $\beta = 200$, the equation is:

$$12 \text{ V} - (V_o + 0.65 \text{ V}) = \frac{(V_o + 12 \text{ V})}{200 \times 1.2 \text{ k}\Omega} \times 120 \text{ k}\Omega$$

$$= (V_o + 12 \text{ V})/2$$

$$(12 - 6 - 0.65) \text{ V} = 3V_o/2$$

$$V_o = 5.35 \text{ V}/1.5 = 3.57 \text{ V}$$

(e)

The input conductance of the circuit is $g_{in} = \dfrac{g_i(g_o + g_E + g_L)}{g_o + g_E + g_L + g_i + g_m}$ (see equation 10A.4 of the textbook (page 455)).

From the data given,

$$g_m = KI_C = 40 \text{ V}^{-1} \times 9.9 \text{ mA} = 39.6 \text{ mA V}^{-1} \,(= 39\,600 \text{ μS})$$

$$g_o = \frac{I_C}{V_A + V_{CE}} = \frac{9.9 \text{ mA}}{(90 + 12) \text{ V}} = 97 \text{ μS}$$

$$g_i = g_m/\beta = 39.6 \text{ mA V}^{-1}/100 = 396 \text{ μS}$$

$$g_E = 1/R_E = 833 \text{ μS}, \quad g_L = 1/R_L = 2000 \text{ μS}$$

Hence, $g_{in} = \dfrac{396 \times 2\,930}{42\,900} \text{ μS} = 27 \text{ μS}$

Solutions

In this calculation, g_o is negligible in the numerator and g_m is by far the largest term in the denominator. In this case therefore, the simpler expression $g_{in} = g_i(g_E + g_L)/g_m = (g_E + g_L)/\beta$ could have been used, and would have given the value 28 µS for the input conductance.

The input resistance is the reciprocal of the input conductance and is therefore 37 kΩ.

The output resistance can be found using equation 10A.9 of the textbook (page 456), since the requirement that g_i, g_E and g_o are much less than g_m is satisfied, as is the requirement that g_o is much less than g_E.

The source resistance is specified as 10 kΩ, so $g_S = 100$ µS, while $g_B = 1/R_B = 8.7$ µS.

So,

$$g_{out} = \frac{(g_S + g_B)g_m + g_i g_E}{g_S + g_B + g_i} = \frac{109 \times 39\,600 + 396 \times 833}{505}\,\text{µS}$$

$$= 9\,200\,\text{µS}$$

The output resistance is therefore ≈ 110 Ω.

Question 10.10

(a)

The total current flowing in R_4 is $I_3 + I_2 = 22$ mA, so $R_4 = 15$ V ÷ 22 mA = 680 Ω.

The voltage across R_3 is the base-emitter voltage of T6, which is very nearly 0.65 V. The current flowing through R_3 is 2 mA, so $R_3 = 0.65$ V ÷ 2 mA = 325 Ω.

The base current of T6 is 20 mA ÷ 100 = 0.2 mA, so the total collector current of T5 is 2.2 mA.

The base current of T5 is therefore 2.2 mA ÷ 200 = 11 µA.

The emitter current of T2 is one-half of the "tail" current and is therefore 100 µA. Since the current gain $\beta = 100$, the collector curent will be 99 µA and the base current will be 0.99 µA.

The current through R_2 is therefore 99 µA − 11 µA = 88 µA.

The voltage across R_2 is the base-emitter voltage of T5 and so $R_2 = 0.65$ V ÷ 88 µA = 7.4 kΩ.

For I_1 to be 200 µA, the current through R_1 must be 200 µA, so

$$R_1 = (30\,\text{V} - 0.65\,\text{V}) \div 200\,\text{µA} = 147\,\text{kΩ}.$$

(b)

(i) *The long-tailed pair.*

The voltage gain of the long-tailed pair is $A_1 = \tfrac{1}{2} \times \dfrac{g_{m2}}{g_{o2} + g_2 + g_{i5}}$,

Analogue transistor circuits

where

$$g_{m2} = KI_{C2} = 40 \text{ V}^{-1} \times 100 \text{ μA} = 4 \text{ mA V}^{-1}$$

$$g_{o2} = \frac{I_{C2}}{V_A + V_{CE2}} = \frac{99 \text{ μA}}{100 \text{ V} + (15 - 0.65 - (-0.65)) \text{ V}} = \frac{99 \text{ μA}}{115 \text{ V}} = 0.86 \text{ μS}$$

$$g_2 = 1/R_2 = 135 \text{ μS}$$

$$g_{i5} = \frac{g_{m5}}{\beta} = \frac{40 \text{ V}^{-1} \times 2 \text{ mA}}{200} = 400 \text{ μS}$$

The voltage gain of the long-tailed pair is therefore:

$$A_1 = \tfrac{1}{2} \times \frac{4 \text{ mA V}^{-1}}{(0.86 + 135 + 400) \text{ μS}} = \frac{2000}{536} = 3.7$$

(ii) *The common-emitter stage.*

The common-emitter stage gain is $A_2 = g_{m5} R_{L(total)} = g_{m5}/g_{L(total)}$ where $g_{L(total)} = g_{in(6)} + g_{o5}$.

$g_{in(6)}$ is the input resitance of the emitter follower stage. For this emitter follower, the resistor R_3 is directly in parallel with the g_i of T6 and therefore results in the equivalent circuit of Figure 10.7 (compare this circuit with the equivalent circuit on page 455 of the textbook).

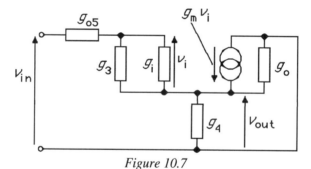

Figure 10.7

The results derived in Appendix 10A of the textbook must therefore be modified by putting $g_B = 0$ and replacing g_i with $(g_i + g_3)$. There is also no load on the emitter follower so $g_L = 0$. The modified form of equation 10A.4 of the textbook is then:

$$g_{in(6)} = \frac{(g_{i6} + g_3)(g_{o6} + g_4)}{g_{o6} + g_4 + (g_{i6} + g_3) + g_{m6}}$$

where $g_{m6} = KI_C = 40 \text{ V}^{-1} \times 20 \text{ mA} = 800 \text{ mA V}^{-1}$, $g_{i6} = g_{m6}/\beta = 8000 \text{ μS}$.

$$g_3 = 1/R_3 = 3080 \text{ μS}, \quad g_{o6} = \frac{I_C}{V_A + V_{CE}} = \frac{20 \text{ mA}}{115 \text{ V}} = 174 \text{ μS} \text{ and } g_4 = 1/R_4 = 1470 \text{ μS}$$

$$g_{in(6)} = \frac{(8\,000 + 3\,080)(174 + 1\,470)}{174 + 1\,470 + (8\,000 + 3\,080) + 800\,000}\,\mu S$$

$$\approx \frac{11\,000 \times 1\,600}{811\,000}\,\mu S$$

$$= 22\,\mu S$$

Now $\quad g_{o5} = \dfrac{I_{C5}}{V_A + V_{CE5}} = \dfrac{2.2\text{ mA}}{60\text{ V} + (15 - 0.65)\text{ V}} = 30\,\mu S$

Finally, $\quad A_2 = \dfrac{g_{m5}}{g_{in(6)} + g_{o5}} = \dfrac{40\text{ V}^{-1} \times 2.2\text{ mA}}{52\,\mu S} = 1700$

(iii) *The emitter follower.*

The gain of the emitter follower is very nearly 1.

The overall amplifier gain is therefore $3.7 \times 1700 = 6\,300$.

(c)

The input resistance of the long-tailed pair is the sum of the input resistances of the two transistors and is therefore equal to $2/g_i$. With 100 µA current in each transistor, $g_m = 4$ mA V^{-1} and $g_i = g_m/\beta = 40\,\mu S$.

The input resistance of each transistor is therefore 25 kΩ and the input resistance of the amplifier is 50 kΩ.

The output resistance of the emitter follower is given by equation 10A.9 of the textbook (page 456) amended to suit the new equivalent circuit (Figure 10.7). The amended equation is:

$$g_{out} = \frac{g_{o5}\,g_{m6} + (g_{i6} + g_3)g_4}{g_{o5} + (g_{i6} + g_3)}$$

where $g_{o5} = 30\,\mu S$, $g_{m6} = 800$ mA V^{-1}, $g_{i6} = 8000\,\mu S$, $g_3 = 3080\,\mu S$ and $g_4 = 1470\,\mu S$.

Hence,

$$g_{out} = \frac{30 \times 800\,000 + 11\,080 \times 1470}{30 + 11\,080} \approx 3600\,\mu S$$

and the output resistance $= r_{out} = 1/(3630\,\mu S) = 275\,\Omega$.

11 BASIC TRANSISTOR DIGITAL CIRCUITS

QUESTIONS

11.1 Figure 11.1 shows the d.c. characteristics of the transistor in the circuit of Figure 11.2. The input voltage can only be either in the band 0 to +0.5 V or in the band +4 V to +5 V.

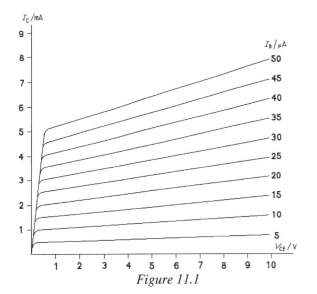

Figure 11.1

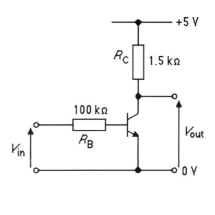

Figure 11.2

(a) By constructing a load line on Figure 11.1, determine whether the output voltage will also be in one or other of these two voltage bands.

(b) If a load resistor of 2.2 kΩ is now connected between the output and the +5 V line, will the output voltages now lie within the two specified voltage bands?

(c) If the output voltage requirement is not met in (b), what changed value of (i) R_C or (ii) R_B will satisfy the output voltage requirement?

11.2 The circuit of Figure 11.2 is now required to have a noise immunity of 0.7 V at the input, so that when the input voltage is between –0.7 V and +1.2 V, the output should be between +4 V and +5 V, and when the input is in the range 3.3 V to 5.7 V, the output should be in the range 0 V to +0.5 V.

(a) Will the original circuit without the load satisfy the requirement? If not, why not?

(b) Is there a combination of values of R_C and R_B which will enable the circuit to satisfy the requirement?

(c) Could the addition of a diode to the circuit enable the requirement to be met, and if so where should the diode be placed? Would a change in the value of R_C or R_B or both also be needed?

Basic transistor digital circuits

11.3 (a) Calculate the collector-emitter voltage of the transistor in Figure 11.3 at 25 °C if $\beta_N = 200$ and $\alpha_I = 0.45$. Assume that the series resistances within the transistor are negligible. Calculate the value of K using the formula $K = q/kT$, where q (the charge on an electron) $= 1.6 \times 10^{-19}$ coulombs, k (Boltzmann's constant) $= 1.38 \times 10^{-23}$ J K^{-1} and T is the temperature in kelvin.

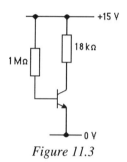

Figure 11.3

(b) Assuming no significant change with temperature in the values of β_N and α_I, what will be the value of V_{CE} at 75 °C?

11.4 The transistor of Question 11.3 has common-base cut-off current $I_{CB0} = 0.2$ nA at 25 °C.

(a) Calculate the value of I_{CE0} at 25 °C.

(b) Calculate the value of I_{CE0} at 75 °C.

11.5 In the circuit of Figure 11.4, the transistor has the following parameters:

transit time τ_t	= 0.25 ns
saturation time τ_s	= 15 ns
emitter capacitance C_e	= 1.2 pF
collector capacitance C_c	= 1 pF
current gain β	= 120
saturation voltage $V_{CE(SAT)}$	= 0.2 V

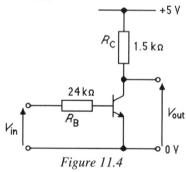

Figure 11.4

The input voltage changes between two levels, +0.2 V and +4.3 V with rise and fall times which are negligibly short. The input voltage source has a resistance of 1 kΩ.

(a) Calculate the four response times t_d, t_f, t_s and t_r and hence calculate t_{on} and t_{off}.

(b) If the base resistor R_B is reduced to 1.5 kΩ, estimate (without performing calculations) the changes which will occur in t_{on} and t_{off}.

(c) Re-calculate all the response times to confirm the validity of your estimate.

Questions

11.6 A capacitor of value 10 pF is connected in parallel with R_B in the circuit of Figure 11.4. Recalculate the switching times t_{on} and t_{off}.

11.7 Figure 11.5 shows the same circuit as in Figure 11.4, but with the transistor replaced by one having an in-built Shottky diode between the base and collector. Assuming that the new transistor has identical characteristics to those of the transistor in Question 11.5, calculate the new value of Q_{BS} and hence of t_s.

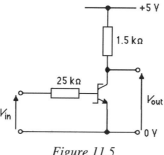

Figure 11.5

11.8 Figure 11.6 shows the characteristic curves of the MOSFET of Figure 11.7. The transistor has the following internal capacitances:

gate-source capacitance C_{gs} 0.01 pF
gate-drain capacitance C_{gd} 0.01 pF
gate-channel capacitance C_{gc} 0.15 pF

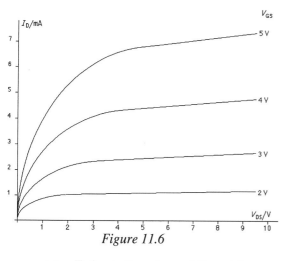

Figure 11.6

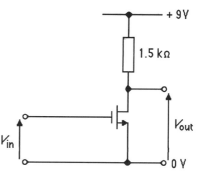

Figure 11.7

(a) Estimate the values of V_T and λ.

(b) If the input voltage switches between 0 V and +5 V, what are the corresponding output voltage levels?

(c) Calculate the charge required to switch the transistor on.

(d) If a turn-on current of 100 µA is available from the source following a positive transition of V_{in}, and a turn-off current of 30 µA following a negative transition of V_{in}, calculate the turn-on and turn-off times of the transistor.

(e) What value of capacitor, attached across the input source instead of the transistor, would take the same charge from the source at each transition?

(f) Figure 11.8 shows one possible source of the input voltage changes. The switch is a very high speed electronic switch having negligible delays. With this input circuit, calculate the rise and fall times (i.e. the 10% to 90% turn-on and turn-off times) for the transistor output voltage.

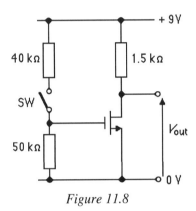

Figure 11.8

SOLUTIONS

Question 11.1

(a)

Figure 11.9 shows, marked (a), the load line for $R_C = 1.5$ kΩ drawn on the transistor characteristics. It joins the points $I_C = 0$, $V_{CE} = 5$ V and $V_{CE} = 0$, $I_C = 5$ V ÷ 1.5 kΩ = 3.3 mA.

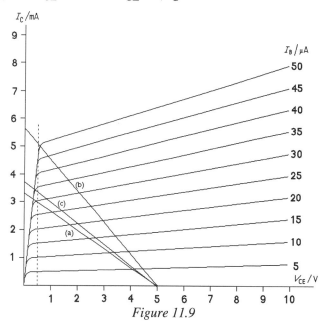

Figure 11.9

With this load line, for the output voltage to be greater than 4 V, the base current must be less than 5 µA. When V_{in} is 0.5 V or less, the base current of the transistor must be negligible, so the output voltage will be +5 V.

When $V_{in} = 4$ V or more, $I_B \geq \dfrac{(4 - 0.65)\text{ V}}{100\text{ k}\Omega} = 33.5$ µA. At this base current, the output voltage is, from the load line, about 0.3 V, which is well within the required 0.5 V.

The circuit therefore meets the output voltage requirements.

(b)

When a load resistor of 2.2 kΩ is connected between the output and the +5 V line, the effective collector resistance is 2.2 kΩ in parallel with 1.5 kΩ, i.e. 890 Ω.

The new load line is shown on Figure 11.9 marked (b). From the load line, it is clear that at a base current of 33.5 µA, the output voltage will be about 1.8 V, which is well outside the required voltage band.

Basic transistor digital circuits

(c)

(i) To meet the requirement, R_C needs to be increased until the effective resistance of R_C and R_L in parallel is sufficiently large to allow a base current of 33.5 µA to result in a collector voltage less than 0.5 V. The mimimum value of collector resistance which will satisfy this requirement is shown by load line (c) on Figure 11.9. This represents a resistance of 5 V ÷ 3.7 mA = 1.35 kΩ

The new required value for R_C is therefore given by;

$$\frac{R_C \times 2.2 \text{ k}\Omega}{R_C + 2.2 \text{ k}\Omega} \geq 1.35 \text{ k}\Omega$$

$$2.2 R_C \geq 1.35 R_C + 2.97 \text{ k}\Omega$$

$$R_C \geq 3.5 \text{ k}\Omega$$

Any value of R_C greater than or equal to 3.5 kΩ will satisfy the requirement when R_B is 100 kΩ.

(ii) For this transistor, any load line which indicates saturation at a base current greater than 45 µA will also show a saturation voltage above 0.5 V. With load line (b), saturation occurs at a base current a little above 50 µA, and at a collector-emitter voltage of about 0.6 V. Any reduction in the value of R_B to increase the base current cannot therefore reduce the saturation voltage to satisfy the output voltage requirements.

Question 11.2

(a)

When the input voltage is +3.3 V, $I_B = \frac{(3.3 - 0.65) \text{ V}}{100 \text{ k}\Omega} = 26.5 \text{ µA}$. At this base current, the output voltage ≈ 0.9 V (see load line (a) on Figure 11.9), which does not satisfy the output voltage requirement.

When the input voltage is +1.2 V, $I_B = \frac{(1.2 - 0.65) \text{ V}}{100 \text{ k}\Omega} = 5.5 \text{ µA}$ and the output voltage is about +4 V, which just satisfies the requirement.

(b)

Any increase in R_C will reduce the slope of the load line, so allowing a lower base current to create saturation in the transistor when the input voltage is high and hence reducing the low output voltage.

Having increased R_C, an increase in R_B to reduce the saturation base current will also reduce the base current when the input voltage is low, so increasing the high output voltage.

For example, if R_C is made 2 kΩ (load line (d) on Figure 11.10), an output voltage less than or equal to 0.5 V can be achieved with a base current of about 22 mA or more. To obtain sufficient base current,

$$R_B \leq \frac{(3.3 - 0.65) \text{ V}}{22 \text{ µA}} = 120 \text{ k}\Omega$$

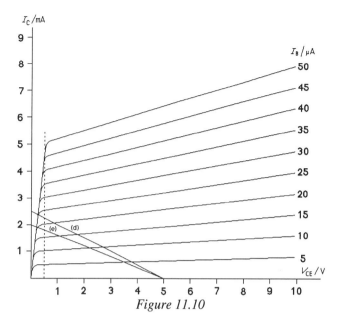

Figure 11.10

However, to satisfy the high ouput voltage condition with this value of R_C, the base current must be less than or equal to about 4 µA, giving:

$$\frac{(1.2-0.65) \text{ V}}{R_B} \leq 4 \text{ µA} \quad or \quad R_B \geq \frac{0.55 \text{ V}}{4 \text{ µA}} = 140 \text{ k}\Omega$$

These two requirements are clearly contradictory, so no suitable value of R_B can be found to satisfy both the high and the low output voltage requirement.

Similar calculations for higher values of R_C will show similar results. For example, for $R_C = 2.5$ kΩ (load line (e) on Figure 11.10), the requirements on R_B become $R_B \leq 150$ kΩ and $R_B \geq 180$ kΩ.

It is therefore reasonable to conclude that there is no suitable combination of values of R_C and R_B which will satisfy the noise margin criterion.

(c)

There are four possible places in the circuit of Figure 11.2 where a diode could be introduced, and they are (i) in the emitter circuit, (ii) in the collector circuit between the +5 V line and the output, (iii) in the collector circuit between the collector and the output and (iv) in series with R_B. (In each case, of course, the diode must be in the correct orientation to allow the circuit currents to flow.)

(i) In the emitter lead of the transistor, the diode would increase the emitter voltage to about +0.6 V and would therefore increase the base voltage to about 1.25 V. The output voltage would also increase by 0.6 V for any given value of V_{CE}. As a result, there would be no possibility of the output voltage falling below about 0.8 V, however large the value of R_C. However, the base current for an input voltage of 1.2 V would fall to a negligible value, so allowing the high output voltage requirement to be met.

Basic transistor digital circuits

(ii) In the collector circuit, between the supply line and the output, the presence of the diode would reduce the collector voltage by 0.6 V for any significant collector current. In effect, the load line is moved to the left by 0.6 V except at extremely low collector currents, giving the load line marked (f) on Figure 11.11 for $R_C = 1.5$ kΩ. An extremely low value of base current (about 1 µA) would be needed to satisfy the high output voltage requirement, and this could only be obtained with a very large value of R_B, which would prevent the achievement of the low output voltage requirement.

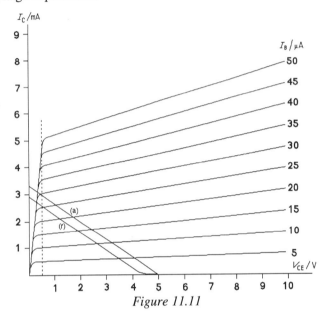

Figure 11.11

(iii) In the collector circuit, between the collector and the output, the diode would make the output voltage 0.6 V more than the collector-emitter voltage at any significant collector current, effectively moving the whole set of transistor characteristics 0.6 V to the right. The output voltage at saturation could then never be less than the 0.6 V of the diode, and with $R_C = 1.5$ kΩ the output voltage at saturation would be about 0.9 V, provided the base current was 32 µA or more.

(iv) A diode in series with the base resistor R_B will add 0.6 V to the base-emitter voltage of the transistor at all but negligible base currents. When the input voltage is 1.2 V, the base current flowing will be negligible, since a voltage of about 1.25 V (0.65 V for the base-emitter junction of the transistor and 0.6 V for the diode) is required to cause any significant current flow. The output voltage will therefore be +5 V. When the input voltage is +3.3 V, the voltage across R_B will be (3.3 − 1.25) V and the value of R_B can be reduced to give a base current of 30 µA or more, which is the current necessary to saturate the transistor when the collector resistor is 1.5 kΩ (see load line (a) on Figure 11.11). The required value of R_B is then (3.3 − 1.25) V ÷ 30 µA = 68 kΩ.

The diode should therefore be placed in series with a reduced value of R_B.

Solutions

Question 11.3

(a)

The base current $I_B = \dfrac{(15 - 0.65)\text{ V}}{1\text{ M}\Omega} = 14.3\text{ μA}$.

The collector current $I_C \approx 15\text{ V}/18\text{ k}\Omega = 0.83\text{ mA}$ so $I_C = 58 I_B$.

If $\alpha_I = 0.45$, $\beta_I = \alpha_I/(1 - \alpha_I) = 0.82$.

At a temperature of 25°C, $K = \dfrac{q}{kT} = \dfrac{1.6 \times 10^{-19}\text{ C}}{1.38 \times 10^{-23}\text{ JK}^{-1} \times (273 + 25)\text{ K}} = 38.9\text{ V}^{-1}$

Substituting these values in equation 11.5 of the textbook (page 464),

$$V_{CE} = \dfrac{1}{K} \ln\left\{\dfrac{(I_B/\alpha_I) + (I_C/\beta_I)}{I_B - (I_C/\beta_N)}\right\}$$

$$= \dfrac{1}{38.9} \ln\left\{\dfrac{I_B/0.45 + 58 I_B/0.82}{I_B - 58 I_B/200}\right\}$$

$$= \dfrac{1}{38.9} \ln\left\{\dfrac{2.2 + 70.7}{1 - 0.29}\right\}\text{ V} = \dfrac{1}{38.9} \ln(102.8)\text{ V} = 0.12\text{ V}$$

(b)

At 75°C, $K = \dfrac{q}{kT} = \dfrac{1.6 \times 10^{-19}\text{ C}}{1.38 \times 10^{-23}\text{ JK}^{-1} \times (273 + 75)\text{ K}} = 33.3\text{ V}^{-1}$

and $V_{CE} = 0.12\text{ V} \times 38.9/33.3 = 0.14\text{ V}$.

Question 11.4

(a)

Using equation 11.4 of the textbook (page 464), $I_{CE0} = (-I_{CS})/(1 - \alpha_N) = I_{CB0}/(1 - \alpha_N)$.

If $\beta_N = 200$, $\alpha_N = \beta_N/(1 + \beta_N) = 200/201 = 0.995$.

So, $I_{CE0} = \dfrac{0.2 \times 10^{-9}\text{ A}}{0.005} = 40\text{ nA}$

(b)

The current I_{CB0} (or I_{C0}) is approximately proportional to n_i^2, where n_i increases by 8% per °C.

Basic transistor digital circuits

Hence, at 75 °C, $\quad I_{CB0} = 0.2 \times 10^{-9} \text{ A} \times \left((1.08)^2\right)^{50} = 0.2 \times 10^{-9} \text{ A} \times 2200 = 0.44 \text{ μA}$

and $\quad I_{CE0} = 200 \times I_{CB0} = 88 \text{ μA}$

Question 11.5

(a)

The currents available to switch the transistor on and off can be found as follows:

$$I_{B(ON)} = (4.3 - 0.65) \text{ V}/(24 + 1) \text{ k}\Omega = 146 \text{ μA}$$

$$I_{B(OFF)} = (0.65 - 0.2) \text{ V}/(24 + 1) \text{ k}\Omega = 18 \text{ μA}$$

The ON collector current of the transistor is:

$$I_{C(ON)} = (5 - 0.2) \text{ V}/1.5 \text{ k}\Omega = 3.2 \text{ mA}.$$

So, $\quad Q_B = I_{C(ON)} \tau_t = 3.2 \text{ mA} \times 0.25 \text{ ns} = 0.8 \text{ pC}$

and

$$Q_{BS} = \tau_S \left(I_{B(ON)} - I_{C(ON)}/\beta\right) = 15 \text{ ns} \times (146 \text{ μA} - 3.2 \text{ mA}/120)$$
$$= 15 \text{ ns} \times 119 \text{ μA}$$
$$= 1.8 \text{ pC}$$

The change in input voltage is from 0.2 V to 4.3 V, so the initial input voltage (and hence the initial base voltage) is 0.2 V and the change in base voltage which must occur to switch the transistor on is $(0.65 \text{ V} - 0.2 \text{ V}) = 0.45 \text{ V}$.

So, the delay time $t_d = \dfrac{(C_c + C_e) \times \Delta V_{BE}}{I_{B(ON)}} = \dfrac{2.2 \text{ pF} \times 0.45 \text{ V}}{146 \text{ μA}} = 6.8 \text{ ns}$

and the fall time $t_f = \dfrac{Q_B + C_c \times \Delta V_{CE}}{I_{B(ON)}} = \dfrac{0.8 \text{ pC} + 1 \text{ pF} \times (5 - 0.2) \text{ V}}{146 \text{ μA}} = 38 \text{ ns}.$

When the input voltage changes from 4.3 V to 0.2 V,

the saturation time $t_s = Q_{BS}/I_{B(OFF)} = 1.8 \text{ pC}/18 \text{ μA} = 100 \text{ ns}$

and the rise time $t_r = \dfrac{Q_B + C_c \times \Delta V_{CE}}{I_{B(OFF)}} = \dfrac{0.8 \text{ pC} + 1 \text{ pF} \times (5 - 0.2) \text{ V}}{18 \text{ μA}} = 310 \text{ ns}$

Hence, $\quad t_{on} = t_d + t_f = 6.8 \text{ ns} + 38 \text{ ns} = 45 \text{ ns}$

$\quad t_{off} = t_s + t_r = 100 \text{ ns} + 310 \text{ ns} = 410 \text{ ns}$

(b)

The reduction of R_B to 1.5 kΩ reduces the total base circuit resistance to 2.5 kΩ. This increases $I_{B(ON)}$ and $I_{B(OFF)}$ by a factor of 10, but also increases Q_{BS} by a factor greater than 10 (since Q_{BS} depends on the difference between $I_{B(ON)}$ and $I_{C(ON)}/\beta$.

t_d, t_f, and t_r will therefore be decreased by a factor of 10, while t_s will increase. The overall effect should be a reduction in t_{on} by a factor of 10 and a reduction in t_{off}, but by a lesser amount.

(c)

As stated above, t_d, t_f, and t_r will decrease by a factor of 10 so the new values will be $t_d = 0.68$ ns, $t_f = 3.8$ ns and $t_r = 31$ ns.

$$Q_{BS} = \tau_s (I_{B(ON)} - I_{C(ON)}/\beta) = 15 \text{ ns} \times (1460 \text{ }\mu\text{A} - 3.2 \text{ mA}/120)$$
$$= 15 \text{ ns} \times 1433 \text{ }\mu\text{A}$$
$$= 21.5 \text{ pC}$$

$$t_s = Q_{BS}/I_{B(OFF)} = 21.5 \text{ pC}/180 \text{ }\mu\text{A} = 119 \text{ ns}$$

Hence,
$$t_{on} = t_d + t_f = 0.67 \text{ ns} + 3.8 \text{ ns} = 4.5 \text{ ns}$$
$$t_{off} = t_s + t_r = 119 \text{ ns} + 31 \text{ ns} = 150 \text{ ns}$$

Question 11.6

As calculated in Question 11.5, $I_{B(ON)} = 146 \text{ }\mu\text{A}$, $I_{B(OFF)} = 18 \text{ }\mu\text{A}$, $I_{C(ON)} = 3.2 \text{ mA}$, $Q_B = 0.8 \text{ pC}$ and $Q_{BS} = 1.8 \text{ pC}$.

From these results, the total charge required to turn on the transistor is given by

$$Q_{on} = (C_c + C_e) \times \Delta V_{BE} + Q_B + C_c \times \Delta V_{CE}$$
$$= 2.2 \text{ pF} \times 0.45 \text{ V} + 0.8 \text{ pC} + 1 \text{ pF} \times 4.8 \text{ V}$$
$$= 6.6 \text{ pC}$$

and the total charge required to turn off the transistor is

$$Q_{off} = Q_{BS} + Q_B + C_c \times \Delta V_{CE}$$
$$= 1.8 \text{ pC} + 0.8 \text{ pC} + 1 \text{ pF} \times 4.8 \text{ V}$$
$$= 7.4 \text{ pC}$$

The speed-up capacitor in the base circuit must be large enough to supply the larger of these charges when the change in input voltage occurs. Since the turn-off charge is the larger, we should check the adequacy of the capacitor for this change before proceeding with the calculation.

Prior to turn-off, the left-hand end of the capacitor will be at +4.3 V while the right-hand end is at +0.65 V. The capacitor voltage is therefore 3.65 V and the stored charge is:

Basic transistor digital circuits

$$Q = CV = 10 \text{ pF} \times 3.65 \text{ V} = 36.5 \text{ pC}.$$

This is much larger than Q_{off}, so the capacitor is sufficiently large.

To determine the actual switch-on and switch-off times, the initial capacitor current at each transient must be calculated.

When the input voltage changes from 0.2 V to 4.3 V (or vice-versa), the initial capacitor current will be equal to the voltage change divided by the resistance in series with the capacitor. This resistance will be simply the 1 kΩ of the source, since the input resistance of the transistor itself is only valid for small signal analysis when the collector current has been established, and we are only concerned with supplying the turn-on charge Q_{on} or the turn-off charge Q_{off}.

So, the initial current is 4.1 V ÷ 1 kΩ = 4.1 mA and the time taken to supply Q_{on} will be

$$t_{on} = Q_{on}/I_{on} = 6.6 \text{ pC}/4.1 \text{ mA} = 1.6 \text{ ns}$$

and

$$t_{off} = Q_{off}/I_{off} = 7.4 \text{ pC}/4.1 \text{ mA} = 1.8 \text{ ns}.$$

Question 11.7

In Question 11.5, the collector-base junction forward bias when the transistor was switched on was (0.65 V − 0.2 V) = 0.45 V.

With the Shottky diode across the collector-base junction, the forward bias is limited to 0.35 V.

The forward bias is reduced by 0.1 V and so KV_D for the collector junction is reduced by the factor 40 V^{-1} × 0.1 V = 4.

Since the charge density next to the collector junction is an exponential function of the junction voltage, the charge density, and hence Q_{BS}, is reduced by the factor $e^4 \approx 55$.

So, Q_{BS} is reduced from 1.8 pC to 1.8 pC / 55 = 0.033 pC.

Consequently, the saturation time t_s becomes

$$t_s = Q_{BS}/I_{B(OFF)} = 0.033 \text{ pC}/18 \text{ mA} = 1.8 \text{ ns}$$

Question 11.8

(a)

The value of λ can be found from the straight portion of any of the characteristic curves, since -1/λ is the point on the V_{DS} axis at which the straight lines, if projected, would meet.

Using the V_{GS} = 5 V line at V_{DS} = 5 V, I_D = 6.8 mA and V_{DS} = 9 V, I_D = 7.3 mA as indicated on Figure 11.12, the slope of the line is:

$$\frac{V_{DS} + 1/\lambda}{I_D} = \frac{5 \text{ V} + 1/\lambda}{6.8 \text{ mA}} = \frac{9 \text{ V} + 1/\lambda}{7.3 \text{ mA}}$$

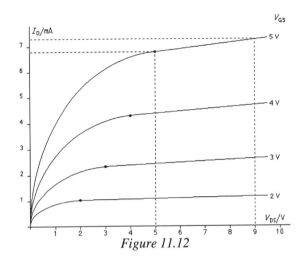
Figure 11.12

from which,

$$1/\lambda(7.3 \text{ mA} - 6.8 \text{ mA}) = (6.8 \text{ mA} \times 9 \text{ V}) - (7.3 \text{ mA} \times 5 \text{ V})$$

$$1/\lambda = 49.4 \text{ V} \approx 50 \text{ V}$$

V_T can be estimated by examining the points at which the characteristic curves change from straight lines (the saturation region) to curves (the linear region). These approximate points are marked on Figure 11.12, and since these points must satisfy the equation $V_{DS} = V_{GS} - V_T$, they show that $V_T \approx 0$ V.

(b)

The load line for the 1.5 kΩ resistor is drawn on the transistor characteristic curves in Figure 11.13.

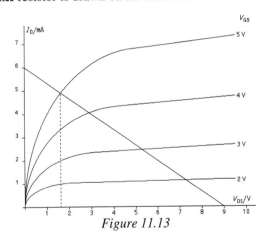
Figure 11.13

When the input is +5 V, the load line shows that the output voltage is +1.6 V. When the input voltage is zero, since the threshold voltage is zero the drain current is zero and the output voltage is +9 V.

Basic transistor digital circuits

(c)

The gate-channel capacitance is normally considered to be one-third in parallel with the gate-drain capacitance and two-thirds in parallel with the gate-source capacitance. The effective gate-drain and gate-source capacitances are therefore:

$$C_{gs(tot)} = C_{gs} + 0.33 C_{gc} = 0.01 \text{ pF} + 0.05 \text{ pF} = 0.06 \text{ pF}$$

$$C_{gd(tot)} = C_{gd} + 0.67 C_{gc} = 0.01 \text{ pF} + 0.10 \text{ pF} = 0.11 \text{ pF}$$

When the transistor turns on, the gate voltage of the transistor changes from 0 to +5 V, while the drain voltage changes from +9 V to +1.6 V, a change of –7.4 V. The gate-source voltage therefore changes by 5 V while the gate-drain voltage changes by 12.4 V.

The charge required from the source to switch the transistor on is therefore:

$$Q_{on} = C_{gs(tot)} \times \Delta V_{gs} + C_{gd(tot)} \times \Delta V_{gd} = (0.06 \times 5 + 0.11 \times 12.4) \text{ pC} = 1.66 \text{ pC}$$

[Since there is no stored saturation charge, and since the threshold voltage is the same as the 'low' input voltage, the charge required to switch the transistor off is also 1.66 pC.]

(d)

With a current of 100 µA available from the source immediately following a positive input voltage transition, the time taken to supply the turn-on charge is

$$t_{on} = 1.66 \text{ pC}/100 \text{ µA} = 16.6 \text{ ns}$$

while the switch-off time will be

$$t_{on} = 1.66 \text{ pC}/30 \text{ µA} = 55 \text{ ns}$$

(e)

Since the switch-on and switch-off charges are the same and equal to 1.66 pC, the equivalent capacitor which would take the same charge from the source at switch-on and switch-off is:

$$C_{EQ} = Q_{on}/\Delta V_g = 1.66 \text{ pC}/5 \text{ V} = 0.33 \text{ pF}$$

(f)

The input voltage circuit of Figure 11.8 can be represented by a Thévenin equivalent circuit for each position of the switch. When the switch is closed (the turn-on state of the circuit), the resistance seen looking back into the source circuit from the transistor gate will be the parallel combination of 40 kΩ and 50 kΩ which is equal to 22 kΩ. The Thévenin equivalent voltage source is clearly +5 V. When the switch is open, there is simply a resistance of 50 kΩ to the 0 V line.

The switch-on condition of the circuit can therefore be represented by the equivalent circuit of Figure 11.14(a), while the switch-off condition can be represented by the equivalent circuit of Figure 11.14(b).

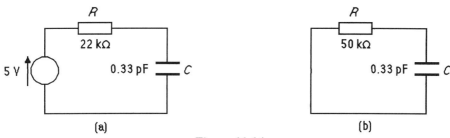

Figure 11.14

The time for the voltage on the capacitor to change from 10% to 90% of its final value is 2.2 *CR* (see the textbook, page 76)

so, rise time = $2.2 \times 22 \text{ k}\Omega \times 0.33 \text{ pF} = 16$ ns

and fall time = $2.2 \times 50 \text{ k}\Omega \times 0.33 \text{ pF} = 36$ ns

12 INTEGRATED-CIRCUIT TECHNOLOGY

This chapter of the textbook has no problem-solving objectives and therefore there are no problems associated with its subject matter.

13 POWER AMPLIFIERS

QUESTIONS

13.1 The biassed double emitter-follower circuit of Figure 13.8 of the textbook (page 537) is required to provide 1W of power to an 8 Ω loudspeaker. The supply voltages are ± 6 V. The transistors can be assumed to have a base-emitter voltage V_{BE} of 0.7 V at their operating currents, while the diodes have a V_D of 0.6 V. The amplifier is to operate in Class A, with the transistor quiescent current being equal to one half of the peak load current.

    (a)   (i)   What is the quiescent transistor current I_Q?

          (ii)   What is the quiescent power dissipation in each transistor?

          (iii)   What is the power dissipated in each transistor at maximum sinusoidal input signal amplitude?

          (iv)   What sinusoidal input voltage amplitude is required to generate the required power output?

    (b)   If the value of the resistors R_E is chosen to minimise distortion due to output resistance variation, what should that value be? What will be the quiescent voltage drop across each emitter resistor?

    (c)   If the current gain β of the transistors is 50, calculate suitable values for R_A and R_B.

    (d)   What will be the amplitude of the current required from the input signal source?

13.2 The same circuit configuration as in Question 13.1 is required to produce a power output of 10 W in an 8 Ω loudspeaker.

    (a)   What is the minimum required value of the supply voltage V_S?

    (b)   The amplifier is given ±16 V supplies, and transistors with parameters: $\beta = 40$, $V_A = 60$ V, $I_{SE} = 10^{-17}$ A, $K = 38.9$ V^{-1}. The diodes have the following parameters: $I_S = 6 \times 10^{-12}$ A, $K = 35$ V^{-1}. The quiescent current is required to be one half of the peak load current, and the value of resistor R_E is chosen to give a voltage drop of 100 mV in the quiescent condition. Calculate suitable values for R_E, R_A and R_B.

    (c)   What is the amplitude of the current taken from the signal source?

13.3 The output transistors of the amplifier of Question 13.2 are each replaced by a 'super β' Darlington configuration as shown in Figure 13.1. The output transistors are the same as were used in Question 13.2, while the driver transistor of each pair has a current gain of 70 and a V_{BE} of 0.7 V. Each biassing diode has been replaced with a pair of diodes in series, so providing better temperature compensation for the two base-emitter voltages in series. Each diode has the same characteristics as those in Question 13.2.

Power amplifiers

(a) Calculate the new values for R_A and R_B.

(b) What is the amplitude of the current which must be provided by the signal source?

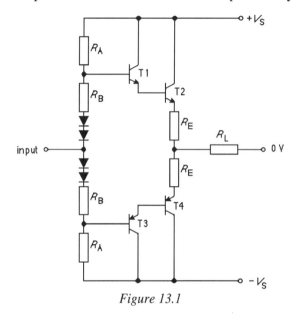

Figure 13.1

13.4 Figure 13.2 shows the configuration of a complementary-symmetry MOSFET output stage in which the load is a 16 Ω loudspeaker.

(a) Assuming that the MOSFETs have adequate current handling capacity and power dissipation, what, typically, will be the maximum available output power?

(b) The MOSFETs have the following characteristics: $\beta = 2$ A V^{-2}, $\lambda = 0.025$ V^{-1}, $V_T = 2$ V (positive for the n-channel device, negative for the p-channel device). The circuit is to be biassed to operate in Class B, with an I_Q of 50 mA. Calculate the required V_{GS} bias voltage. (You will need to use the MOSFET d.c. equation introduced in Chapter 10 of the textbook.)

(c) If the current $I = 1$ mA, select appropriate values for R_A, R_{B1} and R_{B2} to establish the correct quiescent operating conditions, given that the β of T3 and T4 is 200, and that their base-emitter voltages are both 0.65 V.

(d) What, approximately, is the power dissipated in each MOSFET at full output power?

[NOTE: the result of the analysis on page 541 of the textbook cannot be applied to this configuration, since the peak load voltage is considerably less than the supply voltage. Instead, assume that the maximum power dissipation in the transistors occurs when the load power is a maximum.]

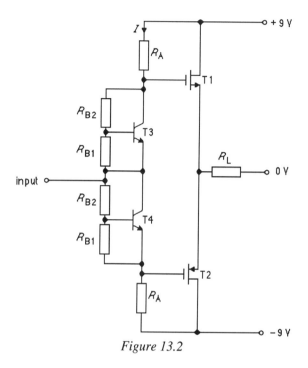

Figure 13.2

13.5 The output stage of an audio amplifier is required to provide 1.5 W of audio frequency power to a loudspeaker from ± 6 V supplies, using output transistors having a maximum collector current of 2 A and maximum power dissipation of 0.6 W. Loudspeakers of 4 Ω, 8 Ω, and 15 Ω impedance are available. It has been decided that the Sziklai configuration of Figure 13.3 will be used, but the amplifier can be biassed to Class A or Class B mode of operation.

(a) Which mode of operation should be used, and which loudspeaker should be chosen? (Assume that $V_{BE3} = V_{BE6} = 0.75$ V and (for this part of the question only) that $V_{BE2} = V_{BE5} = 0.65$ V.)

(b) Assume that the decision has been taken to use Class B operation with a quiescent current of 50 mA. R_E is to be selected to minimize distortion due to variation in output resistance. What should be the value of R_E?

(c) Transistors T1, T2, T4 and T5 all have the following parameters: $V_A = 80$ V, $\beta = 100$, $K = 38.9$ V^{-1}, $I_{SE} = 5 \times 10^{-14}$ A, while transistors T3 and T6 have current gain $\beta = 50$. Choose suitable values for R_A, R_B and R_C.

(d) For your circuit, what is the maximum current which would flow in the output transistors when the load is short-circuited (assuming that the transistors did not blow up before such a current level was reached)? What would then be the power dissipated in each output transistor?

Power amplifiers

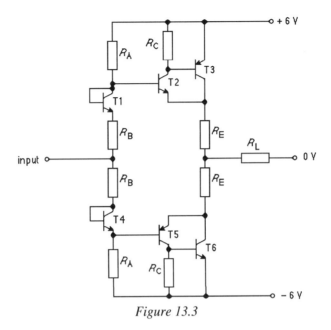

Figure 13.3

13.6 **(a)** The power amplifier of question 13.5 is required to have short-circuit protection of 1.5 A in each output transistor. Sketch the circuit which will give this current protection and calculate the required value of the current sensing resistors, and of any other resistors in your circuit.

(b) An operational amplifier having the configuration shown in Figure 13.11 of the textbook (page 542) is to be used as a first stage amplifier preceding the power amplifier of part (a) so as to give the amplifier an overall gain of 100. The amplifier is to have a passband extending from 30 Hz to 30 kHz (3 dB points). Assuming that the operational amplifier has adequate gain bandwidth product over the passband, make $R_1 = 1$ kΩ and select suitable values for R_3, R_F, C_F, C_1 and C_i. (To start, calculate a value for C_1 such that its reactance, at all frequencies of interest, is much less than the value of R_1, then consider C_1 to be a short circuit while calculating other component values.)

Solutions

13.1

(a)

(i) If the power in the loudspeaker is 1 W, the r.m.s. voltage across the speaker is given by $(V_{r.m.s.})^2/R_L =$ Power, so $(V_{r.m.s.})^2 = 1 \text{ W} \times 8 \text{ }\Omega = 8 \text{ V}^2$ and $V_{r.m.s} = 2\sqrt{2}$ V.

Hence the peak value of a sinusoidal output voltage $= \sqrt{2} \times V_{r.m.s} = 4$ V.

The amplitude of a sinusoidal load current $= 4 \text{ V} \div 8 \text{ }\Omega = 500$ mA.

Hence, the required quiescent current $I_Q = 250$ mA.

(ii) The quiescent power in each transistor $= I_Q \times V_{CE(Q)} = 250 \text{ mA} \times 6 \text{ V} = 1.5$ W.

(iii) At maximum input signal amplitude, V_{CE} varies sinusoidally between $(6-4)$ V and $(6+4)$ V, i.e. between 2 V and 10 V, while I_E varies sinusoidally between 500 mA and 0 mA. The power dissipated is then $V_{average} \times I_{average} = 6 \text{ V} \times 250 \text{ mA} = 1.5$ W.

(iv) Since the gain of an emitter follower is very nearly 1, the input signal amplitude must equal the output signal amplitude, which in this case is 4 V.

(b)

To minimise distortion due to variation in output resistance, the value of R_E should be 0.6 r_{eo}, where r_{eo} is the output resistance of each of the emitter-followers in the quiescent state (see Figure 13.10 of the textbook, page 539).

For an emitter follower driven from an ideal voltage source,

$$r_{eo} = 1/g_m = 1/KI_c = 1/\left(40 \text{ V}^{-1} \times 250 \text{ mA}\right) = 0.1 \text{ }\Omega.$$

Hence, $R_E = 0.06 \text{ }\Omega$ and $V_{RE} = 0.06 \text{ }\Omega \times 250 \text{ mA} = 15$ mV.

(c)

At the positive peak of the input signal, T1 emitter current will be 500 mA, and since $\beta = 50$ the base current will be 10 mA. At the same time, the base voltage of T1 will be 4.75 V (the output voltage plus the V_{BE} of the transistor), so the voltage across $R_A = (6 \text{ V} - 4.75 \text{ V}) = 1.25$ V. Since the base current for T1 must flow through this resistor, the maximum value of R_A is 1.25 V ÷ 10 mA = 125 Ω. To allow for any variation in the current gain of the transistor, a suitable choice for R_A is 100 Ω.

The required value of R_B can now be calculated from the quiescent condition requirements. The quiescent base current is $(250 \text{ mA})/\beta = 5$ mA and the quiescent base voltage is 0.7 V. These values are shown in Figure 13.4, from which,

$$\left(I_D + 5 \text{ mA}\right) \times 100 \text{ }\Omega = (6 \text{ V} - 0.7 \text{ V}) = 5.3 \text{ V}$$

Power amplifiers

$$I_D + 5 \text{ mA} = 53 \text{ mA}$$
$$I_D = 48 \text{ mA}$$

Hence, $R_B = (0.7 \text{ V} - 0.6 \text{ V})/48 \text{ mA} = 2.1 \ \Omega$

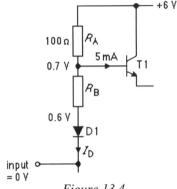

Figure 13.4

(d)

The amplitude of the current required from the source can be found by calculating the difference between the current flowing through D1 and that through D2 at a peak excursion of the input voltage.

At the peak positive excursion of the input, $v_{in} = 4$ V, $I_{D1} = 0.1$ V $\div$ $2.1 \ \Omega = 48$ mA, and since T2 will be cut off, its base current will be zero and the current through D2 will be (6 V + 4 V − 0.6 V) ÷ (100 Ω + 2.1 Ω) = 92 mA.

The peak value of the current from the source will therefore be 92 mA − 48 mA = 44 mA.

Question 13.2

(a)

Using power $= (V_{r.m.s.})^2 / R_L = (V_{peak})^2 / 2R_L$,

$$V_{peak} = \sqrt{\text{power} \times 2R_L}$$
$$= \sqrt{10 \text{ W} \times 16 \ \Omega}$$
$$= 12.75 \text{ V}$$

So the supply voltages must be at least ±13 V.

(b)

At 10 W output power, the peak value of the load current is $V_{peak}/R_L = 12.75 \text{ V}/8 \ \Omega = 1.6$ A, so the required quiescent current I_Q is 0.8 A.

The emitter resistors R_E are required to have a voltage drop of 100 mV, so their value must be 100 mV ÷ 0.8 A = 0.125 Ω.

To calculate the values of R_A and R_B, we need to establish the base-emitter voltage of the transistors, and the forward voltage of the diodes, in the quiescent condition.

The base-emitter voltage of each transistor in the quiescent condition can be found using the equation:

$$I_E = I_{SE}\, e^{KV_{BE}} \times \left(\frac{VA + V_{CE}}{VA}\right)$$

from which,
$$V_{BE} = \frac{1}{K}\ln\left(\frac{I_E}{I_{SE}} \times \frac{VA}{VA + V_{CE}}\right)$$

In the quiescent condition, $I_E = 0.8$ A and $V_{CE} \approx 16$ V. Using the given values of VA, I_{SE} and K,

$$V_{BE} = \frac{1}{38.9}\ln\left(\frac{0.8}{10^{-17}} \times \frac{60}{76}\right) \text{ V} = \frac{1}{38.9}\ln(0.63 \times 10^{17}) \text{ V}$$

$$= \frac{1}{38.9}(\ln 0.63 + 17\ln 10) \text{ V}$$

$$\approx 1.0 \text{ V}$$

This value of the base-emitter voltage will not change significantly when the emitter current doubles to the peak value required by the load.

The diode forward voltage cannot be calculated until we have established the diode quiescent current, which depends on the value of R_A.

To calculate the required value of R_A, we consider the peak input situation.

When the input voltage is at its maximum positive excursion, the voltage at the base of T1 is the load voltage (12.75 V) plus the emitter resistor voltage (0.2 V) plus the base-emitter voltage of the transistor (≈ 1 V). which is 13.95 V.

In this condition, the collector current of T1 is 1.6 A, so the base current is 1.6 A ÷ 40 = 40 mA.

So, the maximum allowable value of $R_A = (16 \text{ V} - 13.95 \text{ V})/40 \text{ mA} = 51$ Ω

Make $R_A = 47$ Ω, and calculate the value of R_B by considering the quiescent condition.

The quiescent values are, $V_{B1} = V_{BE} + I_Q R_E = 1.1$ V, $I_B = 0.8$ A/40 = 20 mA.

If the quiescent diode current is I_D, then

$$(I_D + 20 \text{ mA}) \times 47 \text{ Ω} = 16 \text{ V} - 1.1 \text{ V}$$

$$I_D = \frac{14.9 \text{ V}}{47 \text{ Ω}} - 20 \text{ mA} = 297 \text{ mA}$$

Power amplifiers

We can now calculate the forward voltage of the diode using the equation $I_D \approx I_S e^{KV_D}$ which gives

$$V_D = \frac{1}{K}\ln\left(\frac{I_D}{I_S}\right) = \frac{1}{35}\ln\left(\frac{297 \times 10^{-3}}{6 \times 10^{-12}}\right) \text{ V}$$

$$= \frac{1}{35}\ln(49.5 \times 10^9) \text{ V} = 0.70 \text{ V}$$

So, $\qquad R_B = \dfrac{V_{B1} - V_D}{I_D} = \dfrac{1.1 \text{ V} - 0.70 \text{ V}}{297 \text{ mA}} = 1.35 \, \Omega$

(c)

The amplitude of the current taken from the source is found by subtracting D1 current from D2 current at the positive peak excursion of the input voltage.

Current through D1 = $(V_{B1} - V_{D1}) \div R_B$ = (1.1.V − 0.7 V) ÷ 1.3 Ω = 297 mA.

Current through D2 = (16 V + 12.75 V − 0.6 V) ÷(47 Ω + 1.3 Ω) = 583 mA.

Hence, current required from the source = 583 mA − 297 mA = 286 mA.

Question 13.3

(a)

With the Darlington pair, the quiescent voltage at the base of T1 is

$$V_{B1(Q)} = V_{BE2} + V_{BE1} + I_Q R_E = 1.0 \text{ V} + 0.7 \text{ V} + 0.1 \text{ V} = 1.8 \text{ V}$$

At the maximum positive excursion of the input, the voltage at the base of T1 is

$$V_{B1(peak)} = V_{load} + I_Q R_E + V_{BE2} + V_{BE1} = 12.75 \text{ V} + 0.2 \text{ V} + 1.0 \text{ V} + 0.7 \text{ V} = 14.65 \text{ V}$$

At the same positive excursion, the base current of T1 will be 1.6 A ÷ (40 × 70) = 0.57 mA.

Therefore, the maximum allowable value of $R_A = (16 \text{ V} - 14.65 \text{ V})/(0.57 \text{ mA}) \approx 2.4 \text{ k}\Omega$

Make $R_A = 2 \text{ k}\Omega$ and calculate R_B by considering the quiescent condition.

The quiescent T1 base current is half the peak current = ¹/₂(0.57 mA) = 0.235 mA.

If the current through the biassing diodes is I_D, then,

$$(I_D + 0.235 \text{ mA}) \times 2 \text{ k}\Omega = 16 \text{ V} - 1.8 \text{ V}$$

$$I_D = \frac{14.2 \text{ V}}{2 \text{ k}\Omega} - 0.235 \text{ mA} \approx 6.9 \text{ mA}$$

Solutions

The voltage across each of the two diodes in series will then be

$$V_D = \frac{1}{K}\ln\left(\frac{I_D}{I_S}\right) = \frac{1}{35}\ln\left(\frac{6.9\times 10^{-3}}{6\times 10^{-12}}\right)\text{V} = 0.6\text{ V}$$

so

$$R_B = \frac{V_{B1(Q)} - 2V_D}{I_D} = \frac{1.8\text{ V} - 1.2\text{ V}}{6.9\text{ mA}} = 87\ \Omega$$

(b)

At the peak positive excursion of the input, the current through the upper pair of diodes remains 6.9 mA. The current through the lower diodes is $(12.95 + 16 - 1.2)$ V ÷ 2.09 kΩ = 13.3 mA.

The current taken from the source is therefore 13.3 mA – 6.9 mA = 6.4 mA.

Question 13.4

(a)

Since the gate-source voltage of a MOSFET is typically about 3 V, the output voltage swing will limit at about 3 V less than the supply voltage. Hence the peak value of the load voltage is about 6 V. The maximum output power is therefore

$$P_{o(\max)} = \frac{\left(V_{peak}\right)^2}{2R_L} = \frac{36\text{ V}^2}{32\ \Omega} = 1.1\text{ W}$$

(b)

Using equation 10.7 of the textbook (page 418),

$$I_D = (\beta/2)\times(V_{GS} - V_T)^2(1 + \lambda V_{DS})$$

from which,

$$V_{GS} = \sqrt{\frac{2I_D}{\beta(1+\lambda V_{DS})}} + V_T$$

$$= \sqrt{\frac{2\times 50\text{ mA}}{2\text{ A V}^{-2}\ (1+0.025\times 9)}} + 2\text{ V}$$

$$= 2.2\text{ V}$$

(c)

The actual values of R_{B1} and R_{B2} are not fixed by the information given, although their ratio is important. I will assume that one half of the current I flows through the transistor T3, the other half through resistors R_{B1} and R_{B2}. (You may have chosen a different fraction for the transistor current and hence obtained different, but equally valid, resistor values.)

The collector current of T3 is to be 0.5 mA, so the base current of this transistor will be 2.5 µA (since $\beta = 200$), which is much less than the 0.5 mA flowing through R_{B1} and R_{B2}.

Power amplifiers

Therefore, $V_{CE}/V_{BE} = (R_{B1} + R_{B2})/R_{B1}$ and so $(R_{B1} + R_{B2})/R_{B1} = 2.2\ \text{V}/0.65\ \text{V} = 3.38$.

Also, $R_{B1} + R_{B2} = 2.2\ \text{V}/0.5\ \text{mA} = 4.4\ \text{k}\Omega$

so $R_{B1} = 4.4\ \text{k}\Omega \div 3.38 = 1.3\ \text{k}\Omega$

and hence $R_{B2} = 4.4\ \text{k}\Omega - 1.3\ \text{k}\Omega = 3.1\ \text{k}\Omega$

The resistor R_A, which takes a current of 1 mA, has a voltage drop across it of $(9\ \text{V} - 2.2\ \text{V}) = 6.8\ \text{V}$, so the required value of $R_A = 6.8\ \text{k}\Omega$.

(d)

At maximum load power, each transistor conducts for only one half-cycle of the input waveform, the transistor current varying sinusoidally from the quiescent value of 50 mA to the peak value of 6 V ÷ 16 Ω = 375 mA. At the same time, V_{DS} varies sinusoidally from 9 V to 3 V. So, for this conducting half-cycle,

$$I_{average} = [50 + (375 - 50) \times 2/\pi]\ \text{mA} = 257\ \text{mA}$$

and $$V_{average} = [3 + (9 - 3) \times 2/\pi]\ \text{V} = 6.8\ \text{V}$$

So, the average power in the conducting half-cycle = 257 mA × 6.8 V = 1.75 W.

During the other half-cycle of the input waveform, the current in the transistor decreases rapidly to zero and remains at zero for almost all of the half-cycle, so the power dissipated is negligible. Hence the average power dissipated in one whole cycle, which is the total power dissipated by the transistor, is $1/2 \times 1.75\ \text{W} = 0.875\ \text{W}$.

Question 13.5

(a)

In the Sziklai configuration, the maximum voltage swing at the load is $2V_S - 2V_{BE} = 10.5\ \text{V}$. The peak load voltage is therefore 5.25 V.

Now the maximum load power $= (V_{peak})^2/2R_L$ and the peak load current $= V_{peak}/R_L$.

These values are tabulated below for each available loudspeaker.

impedance (Ω)	Max load power (W)	peak current (A)
15	0.9	0.35
8	1.7	0.66
4	3.4	1.31

From these figures, the 15 Ω loudspeaker is not suitable for the power requirement with the specified supply voltages.

The current carrying capacity of the transistors would allow any of the loudspeakers to be used. However, the power dissipation must also be checked for both Class A and Class B operation.

Solutions

In Class A operation, the maximum power dissipation per transistor is equal to the rated maximum power in the load, so that any of the loudspeakers would cause a transistor overload at maximum power.

In Class B operation, the maximum power dissipated per transistor is ≈ 0.2 times the maximum output power, and would therefore be 0.34 W for the 8 Ω louspeaker and 0.68 W for the 4 Ω loudspeaker. Since the maximum power rating of the transistors is 0.6 W, the only possible selection which satisfies all the criteria is to use the 8 Ω loudspeaker and to use Class B operation of the output stage.

When the amplifier is providing 1.5 W to the loudspeaker, the peak load voltage will be

$$V_{peak} = \sqrt{P_{max} \times 2R_L} = \sqrt{1.5\ \text{W} \times 16\ \Omega} = 4.9\ \text{V}.$$

The peak transistor current will be $V_{peak}/R_L = 4.9\ \text{V}/8\ \Omega \approx 610\ \text{mA}$

(b)

With the Sziklai configuration, to obtain minimum distortion due to output resistance variations, $R_E = 0.12\ r_{eo}$, where $r_{eo} = 1/KI_Q = 1/(40\ \text{V}^{-1} \times 50\ \text{mA}) = 0.5\ \Omega$.

Hence the required value of $R_E = 0.06\ \Omega$.

(c)

R_C is chosen to generate the V_{BE} of T3 at a current which is one tenth of I_Q.

Hence, $\quad R_C = 0.75\ \text{V}/5\ \text{mA} = 150\ \Omega.$

To find the required value of R_A, we must consider the voltage at the base of T2 when the input voltage is at its most positive. The T2 base voltage will be the sum of the peak load voltage, the voltage across R_E and the base-emitter voltage of transistor 2.

The value of the base-emitter voltage can be calculated from the given transistor parameters using the formula:

$$I_E = I_{SE}\, e^{KV_{BE}} \times \frac{VA + V_{CE}}{VA}$$

which re-arranges to give: $\quad V_{BE} = \dfrac{1}{K} \ln\left(\dfrac{I_E}{I_{SE}} \times \dfrac{VA}{VA + V_{CE}}\right)$

In transistor T2, $I_E = [5 + (610 \div 50)]$ mA = 17.2 mA and $V_{CE} = (6 - 0.75 - 4.9)$ V = 0.35 V.

Using the given values of VA, I_{SE} and K,

Power amplifiers

$$V_{BE2} = \frac{1}{38.9} \ln\left(\frac{17.2 \times 10^{-3}}{5 \times 10^{-14}} \times \frac{80}{80.35}\right) V = \frac{1}{38.9} \ln(3.4 \times 10^{11}) V$$

$$= \frac{1}{38.9}(\ln 3.4 + 11 \times \ln 10) V$$

$$\approx 0.68 V.$$

Therefore, at the maximum positive excursion of the input voltage,

$$V_{B2} = V_L + V_{R_E} + V_{BE2} = 4.9 V + (0.61 A \times 0.06 \Omega) + 0.68 V$$

$$= 5.6 V$$

The maximum possible value of R_A is then $(V_S - V_{B2}) \div I_{B2} = 0.4 V \div (17.2 mA / 100) = 2.3 k\Omega$.

To allow for small variations in the current gains of the transistors, it is sensible to make the value a little lower than this maximum, say 2 kΩ.

To find the value of R_B, we must consider the quiescent condition and choose R_B to give the correct base-emitter voltage for T2.

In the quiescent condition, T2 has an emitter current of 5 mA (one tenth of I_Q), and a collector-emitter voltage of (6 V – 0.75 V) = 5.25 V. A repeat of the above calculation for these values gives $V_{BE2} = 0.65$ V, and $V_{B2} \approx 0.65$ V.

The current through R_A is then (6 V – 0.65 V) ÷ 2 kΩ = 2.7 mA. Since the base current of T2 is only 50 µA, the emitter current of T1 is also 2.7 mA.

The diode-connected transistor T1 will have a collector-emitter voltage equal to its base emitter voltage which with this value of emitter current can be calculated to be 0.635 V.

The difference between this voltage and the required base-emitter voltage of T2 is 0.015 V, which is the voltage drop which must exist across R_B.

The required value of R_B is therefore 0.015 V ÷ 2.7 mA = 5.6 Ω.

(In practice, this voltage difference between the base-emitter voltage of T2 and the base-emitter voltage of T1 is so small that a better solution is probably to reduce R_A so that the emitter-base voltages of T1 and T2 are equal and then R_B can be made zero.)

(d)

If the load is short-circuited, the current which will flow in the output transistor (ignoring changes due to temperature increase in the transistors) is limited by the maximum available base current. The maximum current which can flow into the base of T2 (neglecting the voltage drop across R_E and the V_{BE} of T2, both of which would reduce the current somewhat) is 6 V ÷ 2 kΩ = 3 mA. Its collector current would be 100 times bigger, i.e. 300 mA (assuming the transistor is capable of handling such large currents). This would then be approximately the base current of T3 (since only 5 mA flows through R_C) giving a theoretical collector current of T3 of 15 A! At such a current, the power dissipation of T3 would be 90 W and the transistor would act like a fuse, rather than a transistor!

Solutions

Question 13.6

(a)

Figure 13.5 shows the connection of the current-limiting transistors in the circuit. As explained in the textbook, the presence of the current sensing resistor does not affect the performance of the amplifier to any significant effect because it is inside the feedback loop of the two transistors forming the Sziklai configuration. When the voltage drop across the sensing resistor becomes about 0.65 V, T7 (or T8) starts to conduct and takes current out of the base of the driver transistor, so limiting the output current.

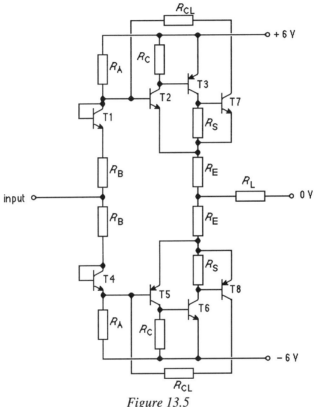

Figure 13.5

To limit the current to 1.5 A, the value of the current sensing resistor must be

$$R_s = 0.65 \text{ V}/1.5 \text{ A} = 0.43 \text{ }\Omega$$

Resistor R_{CL} is included in the collector of T7 to limit the current taken by that transistor. However, its value must be sufficiently low for the transistor to take sufficient current from the base of T2. When the output current is 1.5 A, the base current of T3 is 30 mA, so the collector current of T2 is (30 + 5) mA = 35 mA. The required base current of T2 under current-limit conditions is therefore 350 µA. However, the maximum possible current which can flow through R_A was calculated in Question 13.5(d) to be 3 mA, so the current limiting transistor is required to conduct the difference

Power amplifiers

between these two, i.e. about 2.7 mA. Under these fault conditions, the voltage across R_{CL} will be the V_{BE} of T2 less the V_{CE} of T7, i.e. about 0.45 V if T7 is fully saturated. The maximum value of R_{CL} is therefore $0.45 \text{ V} \div 2.7 \text{ mA} \approx 170 \, \Omega$ and a value of $150 \, \Omega$ would be appropriate.

(b)

Figure 13.6(a) is a functional diagram of the amplifier arrangement.

If the reactance of C_1 is to be much less than the value of R_1 over all frequencies of interest, then, $1/(2\pi f C) \ll 1000 \, \Omega$ when $f = 30$ Hz. So,

$$C \gg \frac{1}{2\pi \times 3 \times 10^4} \text{ F} = 5.3 \, \mu\text{F}$$

If we make $C_1 = 100 \, \mu\text{F}$, this condition will be well satisfied. With this value, the reactance of C_1 is small enough for the presence of C_1 to be ignored during the analysis of the circuit, simplifying the functional diagram to that shown in Figure 13.6(b).

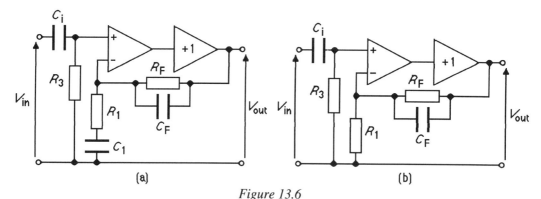

Figure 13.6

The feedback fraction of the amplifier in (b) is:

$$\beta = \frac{R_1}{R_1 + \dfrac{R_F \times 1/j\omega C_F}{R_F + 1/j\omega C_F}} = \frac{R_1}{R_1 + \dfrac{R_F}{j\omega C_F R_F + 1}}$$

$$= \frac{R_1(j\omega C_F R_F + 1)}{R_1(j\omega C_F R_F + 1) + R_F}$$

$$= \frac{R_1(j\omega C_F R_F + 1)}{j\omega C_F R_1 R_F + R_1 + R_F}$$

$$= \frac{R_1}{R_1 + R_F} \times \frac{(1 + j\omega C_F R_F)}{(1 + j\omega C_F R_1 R_F / (R_1 + R_F))}$$

The closed-loop gain $G = 1/\beta$ so,

$$G = \frac{R_1 + R_F}{R_1} \times \frac{1 + j\omega T_1}{1 + j\omega T_2}, \text{ where } T_1 = C_F \cdot \frac{R_1 R_F}{R_1 + R_F} \text{ and } T_2 = C_F R_F$$

The Bode gain diagram for **G** is shown in Figure 13.7.

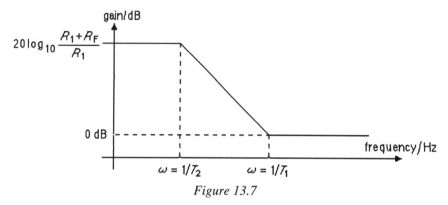

Figure 13.7

The required mid-band gain of 100 is the gain $(R_1 + R_F)/R_1$ and the high frequency 3 dB point of 30 kHz corresponds to the angular frequency $\omega_{c(high)} = 1/T_2$.

Hence, $\quad \dfrac{R_1 + R_F}{R_1} = 100 \text{ and so } R_F = 99 R_1 = 99 \text{ k}\Omega$

and $\quad 30 \text{ kHz} = f_{c(high)} = \omega_c/2\pi = \dfrac{1}{2\pi} \times \dfrac{1}{T_2} = \dfrac{1}{2\pi C_F R_F}$

from which, $\quad C_F = \dfrac{1}{2\pi \times 99 \times 10^3 \ \Omega \times 3 \times 10^4 \text{ Hz}} = 54 \text{ pF}$

The high-pass network of C_i and R_3 must provide the low-frequency 3 dB point at 30 Hz.

R_3 must be chosen to make the resistances of the amplifier input circuits the same, so that the input offset voltage caused by the input bias currents is zero. Because of the presence of C_1 in series with R_1, the resistance to earth of the inverting input circuit is just R_F, so

$$R_3 = R_F = 99 \text{ k}\Omega$$

Also, $\quad 30 \text{ Hz} = f_{c(low)} = \dfrac{1}{2\pi C_i R_3}$

so $\quad C_i = \dfrac{1}{2\pi \times 30 \times 99 \times 10^3} \text{ F} = 54 \text{ nF}$

14 REGULATED D.C. POWER SUPPLIES

QUESTIONS

14.1 Figure 14.1 shows a simple d.c. power supply feeding a 250 load.

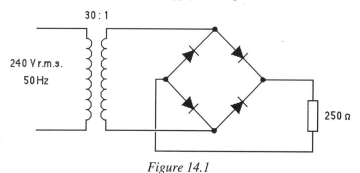

Figure 14.1

(a) What is the d.c. output voltage?

(b) What is the peak-to-peak value of the ripple voltage?

(c) A smoothing capacitor of 4 mF is placed across the load. What is the new peak-to-peak value of the ripple voltage?

(d) What is the new value of the d.c. output voltage?

(e) With the smoothing capacitor present, what is the magnitude of the current peaks in the transformer if each diode conducts for 1 ms in each cycle?

14.2 Design an unregulated power supply which takes its power from the 240 V, 50 Hz mains supply and produces + and − 50 V d.c. The rectifying arrangement is to be that shown in Figure 14.6 of the textbook (page 558) using capacitive smoothing. The current capacity of each supply is to be 25 mA and the peak-to-peak ripple voltage is to be 2 V or less. Specify the required transformer turns ratio, the peak current handling capacity required of the transformer secondary windings and diodes, and the required smoothing capacitors. [You will need to use the information given about the ripple to calculate the approximate time for which each diode conducts.]

14.3 A voltage reference source consists of a combined Zener diode and diode, plus a resistor (Figure 14.2(a)), in series across a full-wave rectifed, capacitor smoothed, d.c supply of 10 V, having 1 V p-p ripple. The combined Zener diode and diode has a voltage drop of 7 V and an internal resistance r_z of 6.5 Ω.

(a) What is (i) the Zener diode current and (ii) the power dissipated in the combined Zener diode and diode?

(b) What is the peak-to-peak ripple voltage on the output?

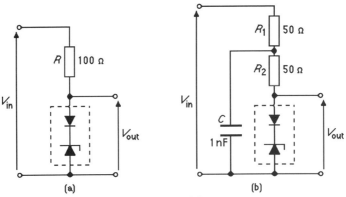

Figure 14.2

(c) The circuit is now modified as shown in Figure 14.2(b). Using the approximate equation (equation 14.1 of the textbook)

$$V_{out} = V_{in} \times \frac{1}{\omega C R_1} \times \frac{r_z}{R_2 + r_z},$$

calculate the new peak-to-peak ripple on the output.

(d) Obtain the equation relating V_{out} to V_{in} in the circuit of Figure 14.3. Use your result to re-calculate the p-p ripple voltage on the output of Figure 14.2(b). Comment on the validity of the approximate equation used in part (c).

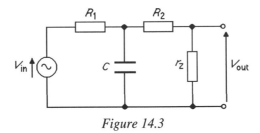

Figure 14.3

(e) Repeat the calculations of parts (c) and (d) using $C = 100$ μF, and again comment on the validity of the approximate equation.

14.4 (a) What value of R_2 is required to make the circuit of Figure 14.4 a bandgap reference voltage source with zero temperature coefficient?

(b) If R_2 is made 75 Ω, what is the value of V_{REF} and what is its temperature coefficient?

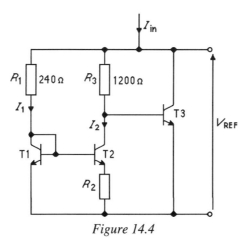

Figure 14.4

14.5 Figure 14.5 is the circuit of a d.c. power supply specified as capable of supplying current in the range 40 mA to 400 mA.

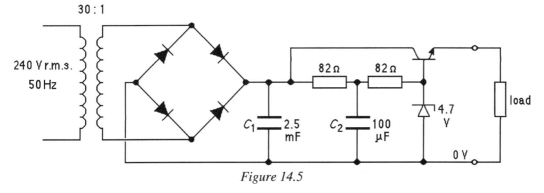

Figure 14.5

(a) What is the peak-to-peak ripple voltage across C_1 at maximum load current, and what is then the peak current taken from the transformer?

(b) If the Zener diode has an internal resistance of 9 Ω, what is the peak-to-peak ripple on the output?

(c) What is the line regulation for a 10% change in supply voltage?

(d) Assuming that the output transistor has a current gain of 50, what is the load regulation?

(e) What is the maximum power dissipated in the output transistor?

(f) If the transistor, plus its heat sink and mounting, has a thermal resistance of 25 °C W^{-1}, and the maximum allowable junction temperature is 100 °C, what is the maximum ambient temerature in which the power supply should be operated?

Regulated d.c. power supplies

(g) Assuming that the output transistor still has a current gain of 50 at very high currents, what will be the short-circuit current, neglecting the effects of a rise in the transistor temperature (and assuming that the transistor does not blow up). What will be the power dissipated in the output transistor under these conditions?

14.6 You are required to design a power supply using the configuration of Figure 14.6.

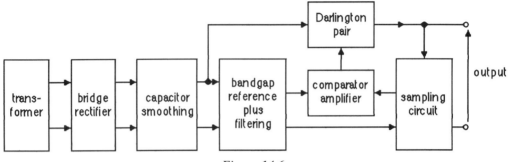

Figure 14.6

The supply should meet the following specification.

Output voltage range	2 V to 20 V.
Output current range	50 mA to 1 A.
Output ripple	< 10 mV.
Load regulation (min. to full load)	0.5%.
Line regulation (10% line voltage change)	0.5%.
Overload protection current	2 A.
Usable temperature range	0 °C to 40 °C.

Sketch the circuit of the power supply and specify the value of all circuit components. Specify the required transformer turns ratio and the maximum peak current requirement of the transformer and the diodes. Specify the required current, voltage and power handling capacity of the Darlington pair output transistors. Specify the minimum gain requirement for the comparator amplifier.

The bandgap voltage reference requires a minimum current of 10 mA and has an internal resistance of 5 Ω. You may assume that an output transistor for the Darlington pair is available having a current gain of 30, and a driver transistor having a current gain of 50. The thermal resistance of the output transistor, together with its mounting (but not its heat sink) is 0.7 °C W^{-1} and the maximum operating temperature of the silicon junctions of the output transistor should be assumed to be 120 °C.

[**Hint**. The minimum gain requirement of the comparator amplifier can be determined from the load regulation requirement and the effect of feedback on the reduction in output voltage change. Don't forget to include the sensing resistor for the current protection circuit when calculating the load regulation.]

Solutions

SOLUTIONS

Question 14.1

(a)

With a turns ratio of 30:1, the transformer secondary voltage is 240 V ÷ 30 = 8 V r.m.s. The peak secondary voltage is therefore 8 V × $\sqrt{2}$ = 11.3 V.

The voltage drop across two conducting diodes in series is about 1.3 V so the peak value of the load voltage = 10 V.

The bridge rectifier produces full-wave rectification, so the average value of the output is $2/\pi$ times the peak value.

Hence the average value of the output, which is the d.c. level, is $20/\pi$ V = 6.4 V

(b)

The peak-to-peak ripple voltage = 10 V because there is no smoothing of the full-wave rectified output.

(c)

With the 4 mF smoothing capacitor, the discharge time constant is 4×10^{-3} F × 250 Ω = 1 s. This is much greater than the discharge time of 10 ms, so the rate of discharge can be considered to be constant throughout the discharge. The initial rate of discharge of the capacitor is V/CR = 10 V ÷ 1 s = 10 V s^{-1}. This rate of discharge lasts for approximately 10 ms, so the voltage fall is 100 mV. This is the peak-to peak value of the ripple voltage.

(d)

Since the output voltage varies linearly from 10 V to 9.9 V in each half-cycle, the d.c output is 9.95 V.

(e)

Since each diode conducts for 1 ms in each cycle, the capacitor discharges for the remainder of each half-cycle (remember, this circuit is a full-wave rectifier), i.e. for 9 ms.

The current flowing in the load ≈ 10 V ÷ 250 Ω = 40 mA, so the charge lost from the capacitor during the discharge is 40 mA × 9 ms = 360 μC.

The charge gained in each half-cycle must therefore also be 360 μC, and since the charge lasts for 1 ms, the average current during the charging must be 360 mA. Since the current pulse has a roughly triangular shape (see Figure 14.8 of the textbook), the peak value of the current will be about twice the average, or 720 mA. (This will be an overestimate of the peak value, but that is an error on the safe side when trying to determine the transformer current requirements.)

Regulated d.c. power supplies

Question 14.2

Figure 14.7 shows the arrangement of the power supply.

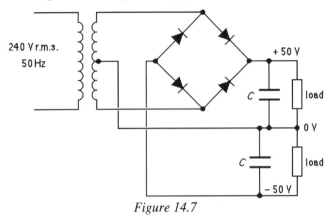

Figure 14.7

With 50 V d.c. output and 2 V peak-to-peak ripple, the output varies between 51 V and 49 V. Since only one conducting diode is in series with a load at any time, the transformer secondary peak voltage required is approximately 51.7 V.

The peak input voltage to the transformer is 240 V × √2 = 340 V, so the turns ratio required for each half of the transformer = 340 V ÷ 51.7 V ≈ 6.58:1.

[A fractional turns ratio is perfectly acceptable, the primary can have a number of turns which is a multiple of 658, and each half of the secondary would then have 100 turns for each 658 on the primary.]

The load current will be 25 mA, so the load resistance is 50 V ÷ 25 mA = 2 kΩ.

If the ripple is 2 V peak-to-peak, the rate of change of output voltage during capacitor discharge (assuming a linear decrease in voltage) ≈ 2 V ÷ 10 ms = 200 V s^{-1}.

However, the initial rate of fall of capacitor voltage in a discharging CR circuit is V/CR, so $V/CR = 200$ V s^{-1}. Hence,

$$C = \frac{51 \text{ V}}{2 \text{ k}\Omega \times 200 \text{ V s}^{-1}} \approx 130 \text{ μF}$$

With 2 V peak-to-peak ripple, the capacitor starts to charge when the transformer voltage reaches 49 V, and continues charging until the voltage reaches 51 V.

If the input voltage is represented by $v = V\sin\theta$, the charging takes place from $\theta = \sin^{-1} 49/51 = 74°$ to $\theta = \sin^{-1} 51/51 = 90°$.

So, the charging occurs for 16° out of each 180° of the sinewave voltage. Hence, the charging time = 16/180 × 10 ms = 0.9 ms and the discharge time = 9.1 ms.

The charge lost by each capacitor in each half-cycle = $I \times t$ = 25 mA × 9.1 ms = 227 µC.

The charge to be gained by each capacitor is also 227 µC, so the average current during charging = 227 µC ÷ 0.9 ms = 252 mA.

The peak current is therefore about twice this value or approximately 500 mA.

The maximum reverse voltage across any diode will occur when the outputs are at their peak value of 51 V. Consider the half-cycle when the 'top' end of the transformer is at +51.7 V and the 'bottom' end at −51.7 V. The bottom right diode and the top left diode will then be reverse biassed to a voltage of 102.7 V. Similarly, at the other peak of the transformer output voltage, the top right and bottom left diodes will be reverse biassed to a voltage of 102.7 V.

Question 14.3

(a)

(i) Since the voltage drop across the Zener diode plus diode is 7 V, the voltage drop across R is 3 V and the current through R is 30 mA.

(ii) The power dissipated in the Zener diode plus diode = 30 mA × 7 V = 210 mW.

(b)

Using the internal resistance to represent the Zener plus diode, the equivalent circuit for changes in voltage is shown in Figure 14.8.

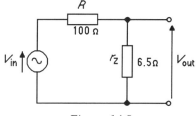

Figure 14.8

From the figure, the output ripple will be given by:

$$V_{out} = \frac{V_{in} r_z}{R + r_z} = \frac{1 \text{ V} \times 6.5}{106.5} = 61 \text{ mV}$$

(c)

Because the input to the voltage reference circuit is from a full-wave rectifier, the fundamental frequency of the ripple voltage will be 100 Hz. Hence $\omega = 2\pi f = 630$ rad s^{-1}.

Hence,

$$V_{out} = V_{in} \times \frac{1}{\omega C R_1} \times \frac{r_z}{R_2 + r_z} = 1 \text{ V} \times \frac{1}{630 \text{ rad s}^{-1} \times 10^{-3} \text{ F} \times 50 \text{ }\Omega} \times \frac{6.5 \text{ }\Omega}{56.5 \text{ }\Omega} = 3.65 \text{ mV}$$

Regulated d.c. power supplies

(d)

Figure 14.9 below is a repeat of Figure 14.3, and I have labelled the voltage V_1 at the junction of R_1 and R_2.

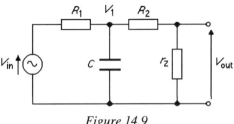

Figure 14.9

$$\frac{V_{out}}{V_1} = \frac{r_z}{R_2 + r_z}$$

$$\frac{V_1}{V_{in}} = \frac{Z}{R_1 + Z} \quad \text{where } Z \text{ is } C \text{ in parallel with } R_2 + r_z.$$

so,

$$Z = \frac{\frac{1}{j\omega C} \times (R_2 + r_z)}{\frac{1}{j\omega C} + (R_2 + r_z)} = \frac{R_2 + r_z}{1 + j\omega C(R_2 + r_z)},$$

and

$$\frac{V_1}{V_{in}} = \frac{(R_2 + r_z)/(1 + j\omega C(R_2 + r_z))}{R_1 + \left[(R_2 + r_z)/(1 + j\omega C(R_2 + r_z))\right]}$$

$$= \frac{R_2 + r_z}{R_1 + j\omega C R_1(R_2 + r_z) + R_2 + r_z}$$

$$= \frac{R_2 + r_z}{R_1 + R_2 + r_z} \times \frac{1}{j\omega C R_{EQ} + 1} \quad \text{where } R_{EQ} = \frac{R_1 \cdot (R_2 + r_z)}{R_1 + (R_2 + r_z)}$$

Therefore,

$$\frac{V_{out}}{V_{in}} = \frac{V_{out}}{V_1} \times \frac{V_1}{V_{in}} = \frac{r_z}{(R_2 + r_z)} \times \frac{(R_2 + r_z)}{(R_1 + R_2 + r_z)} \times \frac{1}{j\omega C R_{EQ} + 1}$$

$$= \frac{r_z}{(R_1 + R_2 + r_z)} \times \frac{1}{j\omega C R_{EQ} + 1}$$

For the component values of the circuit of Figure 14.2(b),

$$CR_{EQ} = \frac{CR_1(R_2 + r_z)}{R_1 + R_2 + r_z} = \frac{10^{-3} \times 50 \times 56.5}{106.5} \text{ s} = 26.5 \text{ ms}$$

so,
$$\frac{V_{out}}{V_{in}} = \frac{6.5}{106.5} \times \frac{1}{j630 \times 26.5 \times 10^{-3} + 1} = \frac{0.061}{j16.7 + 1}$$

$$\left|\frac{V_{out}}{V_{in}}\right| = \frac{0.061}{\sqrt{16.7^2 + 1}} = 3.65 \times 10^{-3}$$

If $|V_{in}| = 1$ V then $|V_{out}| = 3.65$ mV, which is the same result as was obtained in part (c) using the approximate equation. It is clear that for this example, the approximate equation gives an accurate result.

(e)

If C is changed from 1 nF to 100 μF, then, using the approximate equation,

$$V_{out} = V_{in} \times \frac{1}{\omega CR_1} \times \frac{r_z}{R_2 + r_z} = 1\text{ V} \times \frac{1}{630 \times 10^{-4} \times 50} \times \frac{6.5}{56.5} = 36.5\text{ mV}$$

Using the exact equation,

$$CR_{EQ} = \frac{CR_1(R_2 + r_z)}{R_1 + R_2 + r_z} = \frac{10^{-4} \times 50 \times 56.5}{106.5}\text{ s} = 2.65\text{ ms}$$

so,
$$\frac{V_{out}}{V_{in}} = \frac{6.5}{106.5} \times \frac{1}{j630 \times 2.65 \times 10^{-3} + 1} = \frac{0.061}{j1.67 + 1}$$

$$\left|\frac{V_{out}}{V_{in}}\right| = \frac{0.061}{\sqrt{1.67^2 + 1}} = 31.3 \times 10^{-3}$$

If $|V_{in}| = 1$ V then $|V_{out}| = 31.3$ mV, which differs from the result obtained using the approximate equation by about 17%.

For this example, the approximate equation gives the correct order of magnitude, but not an accurate value for the output ripple. This is because the reactance of C at 100 Hz (≈ 16 Ω) is not very much less than the resistance of resistors R_1 and R_2, and so the second part of the network of Figure 14.9 loads the first part of the network to an appreciable extent.

Question 14.4

(a)

The bandgap voltage reference source has an output voltage $V_{REF} = 1.236$ V and zero temperature coefficient.

If the circuit of Figure 14.4 is a bandgap reference source, the voltage across R_3 will be 1.236 V − 0.65 V = 0.586 V, and I_2 will be 0.586 V ÷ 1200 Ω = 488 μA.

Since the voltage across the base-emitter junction of T1 will be almost equal to the base-emitter voltage of T3, the voltage across R_1 will also be 0.586 V and the current I_1 through it will be 0.586 V ÷ 240 Ω = 2.44 mA.

The ratio of I_1 to I_2 is therefore 5, so

$$V_{BE1} - V_{BE2} = \frac{\ln(5)}{40 \text{ V}^{-1}} = 40.2 \text{ mV} \quad \text{(see equation 14.5 of the textbook)}.$$

The voltage across R_2 must therefore be 40.2 mV and so

$$R_2 = \frac{40.2 \text{ mV}}{488 \text{ μA}} = 82.5 \text{ Ω}$$

We can check that the reference voltage has zero temperature coefficient as follows.

Using equation 14.3 of the textbook,

$$\Delta V_{BE1} - \Delta V_{BE2} = \frac{V_{BE1} - V_{BE2}}{T} = \frac{40.2 \text{ mV}}{293 \text{ K}} \quad \text{(at 20 °C)}$$

and using equation 14.6 of the textbook, the change in I_2 per degree rise in temperature is

$$\Delta I_2 = \frac{\Delta V_{BE1} - \Delta V_{BE2}}{R_2} = \frac{40.2 \text{ mV}}{293 \text{ K}} \times \frac{1}{82.5 \text{ Ω}} = 1.66 \text{ μA}$$

The voltage change across R_3 is therefore 1.66 μA K^{-1} × 1200 Ω = 1.996 mV K^{-1}.

Since the temerature coefficient of the base-emitter junction voltage of T3 is −2 mV K^{-1}, the temperature coefficient of the reference voltage, which is the sum of these two voltages, is almost exactly zero. (The difference is due to rounding errors in the maths, the temperature coefficient will be exactly zero if the output voltage is 1.236 V.)

(b)

If R_2 is made 75 Ω instead of 82.4 Ω, I_1 will still be 5 times I_2 because the voltage drops across R_1 and R_3 remain equal. The voltage across R_2 therefore remains 40.2 mV, so I_2 must increase.

The new value of I_2 = 40.2 mV ÷ 75 Ω = 536 μA and the voltage drop across R_3 becomes 536 μA × 1200 Ω = 643 mV.

The output voltage of the circuit therefore becomes (0.643 + 0.65) V = 1.293 V.

Now $\Delta I_2 = \dfrac{40.2 \text{ mV}}{293 \text{ K} \times 75 \text{ Ω}} = 1.83 \text{ μA K}^{-1}$ and $\Delta V_{R_3} = 1.83 \text{ μA K}^{-1} \times 1200 \text{ Ω} = 2.196 \text{ mV K}^{-1}$

The temperature coefficient of the output voltage is therefore

$$\Delta V_{\text{out}} = (2.196 - 2) \text{ mV K}^{-1} \approx 0.2 \text{ mV K}^{-1}.$$

Question 14.5

(a)

The peak value of the transformer output voltage = 240 V × √2 ÷ 30 = 11.3 V.

After full-wave rectification, the peak value will be 11.3 V – 2 × 0.65 V = 10 V.

The maximum current taken by the load = 400 mA, so the current taken from capacitor C_1 = 400 mA – (10 V – 4.7 V) ÷ 164 Ω ≈ 430 mA.

This is effectively a resistive load on C_1 of R_{eff} = 10 V ÷ 430 mA = 23 Ω.

The discharge time constant = $C_1 R_{eff}$ = 2.5 × 10⁻³ F × 23 Ω = 57.5 ms.

This time constant is much longer than the discharge time (≈ 10 ms), so the discharge will be almost linear at a rate

$$\frac{V}{C_1 R_{eff}} = \frac{10 \text{ V}}{57.5 \text{ ms}} = 174 \text{ V s}^{-1}.$$

In 10 ms, the voltage falls 1.74 V, so the p-p ripple voltage across capacitor C_1 is 1.74 V.

To find the peak current in the transformer secondary winding, we need to know the time for which the capacitor is charging. As in the solution to Question 14.2, we can calculate that time as follows.

Representing the sinewave by $v = V\sin \theta$ then when v = (10 – 1.74) V, θ = sin⁻¹ 8.26/10 ≈ 56°, which represents a charging time of (90° – 56°) ÷ 180° × 10 ms ≈ 1.9 ms, and consequently a discharge time of 8.1 ms (not 10 ms as previously assumed).

[This new value of discharge time can be used to re-calculate the ripple amplitude and so re-calculate the angle θ and the charging time. You could continue this iteration of the calculation until the exactly the same result was obtained on two successive iterations, but this is far too tedious and unnecessary for the accuracy required in this type of calculation. However, two iterations produce the more accurate estimate that the ripple amplitude is about 1.5 V and the charging time is 1.7 ms, which are the figures I shall use.]

The charge lost from C_1 in 8.3 ms = 440 mA × 8.3 ms = 3.7 mC, so the average charging current during the 1.7 ms charging time = 3.7 mC ÷ 1.7 ms ≈ 2.2 A. This means that the peak charging current will be a little less than twice this figure, or about 4 A.

(b)

The equivalent circuit for the ripple calculation is shown in Figure 14.10.

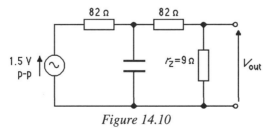

Figure 14.10

Regulated d.c. power supplies

At 100 Hz, the reactance of the capacitor is given by $X_C = \dfrac{1}{2\pi f C} = \dfrac{1}{2\pi \times 100 \times 10^{-4}} \Omega = 16\, \Omega$. This is considerably smaller than 82 Ω, so the approximate formula (equation 14.1 of the textbook) can be used. A more accurate value could be obtained by using the equation developed as part of Question 14.3.

Using the approximate equation,

$$V_{out} = \dfrac{V_{in}}{\omega C R_1} \times \dfrac{r_z}{R_2 + r_z} = \dfrac{1.5}{630 \times 10^{-4} \times 82} \times \dfrac{9}{82 + 9}\, V \approx 29\, mV$$

The ripple across the Zener diode is 29 mV, and since the base-emitter voltage of the output transistor does not change to any significant extent with small changes in output current, the output ripple at maximum current is also 29 mV.

(c)

To find the line regulation, we need to determine by how much the Zener voltage increases with a 10% increase in the supply voltage.

If the supply voltage increases by 10%, the voltage across C_1 will increase by 10%, i.e. by 1 V, and the voltage across the Zener diode, and hence the output voltage, will increase by

$$1\, V \times 9\, \Omega/(164 + 9)\, \Omega = 52\, mV.$$

Hence, line regulation $= \dfrac{52\, mV}{4\, V} \times 100\% = 1.3\%$.

(d)

To find the load regulation, we need to know the change in output voltage as the load current changes from its minimum to its maximum value. This change will be made up of any change in the Zener voltage, combined with any change in the base-emitter voltage of the output transistor.

As the output current increases from 40 to 400 mA, the change in V_{BE} can be calculated using equation 9.14 of the textbook:

$$I_C = I_{SE}\, e^{KV_{BE}} \times \dfrac{VA + V_{CE}}{VA},$$

from which

$$V_{BE} = \dfrac{1}{K} \ln\left(\dfrac{I_C}{I_{SE}} \times \dfrac{VA}{VA + V_{CE}}\right) = \dfrac{1}{K}\ln(I_C) - \dfrac{1}{K}\ln(I_{SE}) + \dfrac{1}{K}\ln\left(\dfrac{VA}{VA + V_{CE}}\right)$$

Since V_{CE} changes very little as the load current changes from 40 mA to 400 mA, the last term will be the same for both values of current, giving:

$$\Delta V_{BE} = V_{BE}(1) - V_{BE}(2) = \dfrac{1}{K}\ln(0.4) - \dfrac{1}{K}\ln(0.04) = \dfrac{1}{K}\ln\left(\dfrac{0.4}{0.04}\right) = \dfrac{1}{K}\ln(10).$$

So, $\Delta V_{BE} = \dfrac{1}{40 \text{ V}^{-1}} \ln(10) \approx 58 \text{ mV}$.

This change is an increase in V_{BE} with increase of current and therefore represents a fall in the output voltage.

When the load current increases from 40 mA to 400 mA, the base current must increase from 0.8 mA to 8 mA, a rise of 7.2 mA. This must be accompanied by a reduction in the Zener current of 7.2 mA, which causes a decrease in Zener voltage of 7.2 mA × 9 Ω = 65 mV.

Also, when the load current is only 40 mA, the ripple on C_1 will be less than the ripple at 400 mA by a factor of 10. The average voltage on C_1 will be larger, being 10 V – 0.075 V (half the ripple) instead of 10 V – 0.75 V at 400 mA, i.e. very nearly 10 V instead of 9.25 V. Thus an *increase* in load current from 40 to 400 mA will result in a decrease in Zener voltage of, from Figure 14.10,

$$\Delta V_z = 0.75 \text{ V} \times \dfrac{9 \text{ }\Omega}{164 \text{ }\Omega + 9 \text{ }\Omega} = 39 \text{ mV}$$

The total change in output voltage will therefore be 39 mV + 65 mV + 58 mV = 162 mV.

The load regulation is then $= \dfrac{162 \text{ mV}}{4 \text{ V}} \times 100\% = 4\%$.

(e)

The maximum output transistor power dissipation occurs at maximum load current and is:

$$P = V_{CE} I_E = (10 - 4) \text{ V} \times 400 \text{ mA} = 2.4 \text{ W}.$$

(f)

The maximum power dissipation in the output transistor is 2.4 W, so the temperature rise above ambient = 2.4 W × 25 °C W^{-1} = 60 °C.

Since the maximum junction temperature is 100 °C, the maximum ambient temperature = 40 °C.

(g)

With a short-circuited output, the base voltage of the output transistor will fall to about 0.7 V, the Zener diode will stop conducting and all the available current through the two 82 Ω resistors will flow into the transistor base.

The short-circuit base current is therefore (10 V – 0.7 V) ÷ 164 Ω = 56.7 mA, and the short-circuit output current will be 56.7 mA × 50 ≈ 2.8 A.

The power dissipated in the transistor will be ≈ 2.8 A × 10 V = 28 W.

[This calculated power would amost certainly be sufficient to destroy the transistor, unless it had been chosen to handle such a large power dissipation.]

Regulated d.c. power supplies

Question 14.6

As with all design exercises, two people working independently will arrive at different solutions, depending on assumptions made about adequate safety margins etc. You should not therefore expect my solution to be the same as yours, but the overall circuit should be similar, and components should have similar calculated values.

Your circuit configuration should be as shown in Figure 14.11. T1 and T2 form the Darlington pair while T3 and R_S form the overload protection circuit.

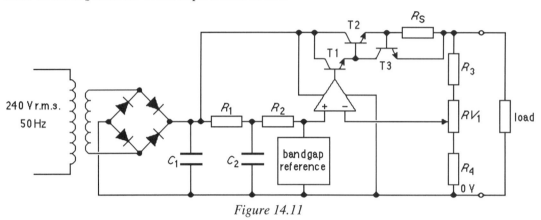

Figure 14.11

Transformer turns ratio

To obtain a 20 V output, the minimum voltage at C_1 should be about 22 V to allow for the voltage drop across the Darlington pair. This minimum voltage should exist even at the lowest point of the ripple voltage.

A reasonable starting point is to assume a maximum voltage of 24 V at C_1, with not more than 2 V p-p ripple. (If necessary, this starting point can be reviewed as the design progresses.)

Allowing for the voltage drop across two diodes, the transformer peak output voltage should be 25.4 V, which gives a turns ratio for the transformer of 240 V × √2 ÷ 25.4 V = 13.4.

Capacitor C_1

The current taken from C_1 will have a maximum value of 1 A (the current through the reference is negligible in comparison) and this must not reduce the capacitor voltage by more than 2 V during one half-cycle.

So, $\quad 2\text{ V} \geq \dfrac{1\text{ A} \times 10\text{ ms}}{C_1} \quad$ and so $\quad C_1 \geq 5 \times 10^3\text{ F} = 5\text{ mF}$.

Normally, C_1 is not made any larger than necessary to minimize the amplitude of the current spikes in the transformer and diodes, so I have chosen $C_1 = 5$ mF.

With this value of capacitor, the average voltage at C_1 will be 23 V and the ripple amplitude will be 2 V p-p.

Resistors R_1 and R_2

The bandgap reference voltage source requires a current of 10 mA at a voltage of 1.236 V, so

$$R_1 + R_2 \le \frac{(23 - 1.236) \text{ V}}{10 \text{ mA}} \approx 2.2 \text{ k}\Omega$$

To give some measure of safety against any reduction in C_1 voltage, make $R_1 = R_2 = 1$ kΩ.

Capacitor C_2

The ripple requirement at the output of the power supply is $\le$ 10 mV p-p, and the worst case for ripple will be when the output voltage is 20 V (because the voltage reference ripple is amplified by 20/1.236 ($\approx$ 16) at that voltage) and when the output current is 1 A (when the ripple at C_1, and hence at the voltage reference, will be largest).

At 20 V output, to obtain an output ripple $\le$ 10 mV, the voltage reference ripple must be less than 10 mV $\div$ 16 $\approx$ 0.6 mV, and when the output current is 1 A, the ripple at C_1 is 2 V.

So, using the equation $\quad V_{out} = \dfrac{V_{in}}{\omega C_2 R_1} \times \dfrac{r_z}{R_2 + r_z}$,

we obtain,

$$C_2 \ge \frac{V_{in}}{V_{out}} \times \frac{1}{\omega R_1} \times \frac{r_z}{R_2 + r_z} = \frac{2 \text{ V}}{0.6 \text{ mV}} \times \frac{1}{630 \times 10^3} \times \frac{5}{1005} = 26 \text{ }\mu\text{F}$$

I will choose to make $C_2 = 30$ µF to provide some margin of safety.

[As a check, at 100 Hz the reactance of a 26 µF capacitor is 61 Ω, which is much less than 1 kΩ, so the use of the approximate equation was justified.]

The sampling circuit

The resistance of the sampling circuit must be much greater than the load resistance, but much less than the input resistance of the comparator amplifier. As we have seen previously, the input resistance of an op-amp can be very large, while the load resistance must be in the range 2 Ω (1 A at 2 V) and 400 Ω (50 mA at 20 V). A reasonable starting point for the selection of the resistor values for the sampling circuit is to say that the total resistance should be of the order of 10 kΩ.

R_3 is present to limit the available output voltage at the low end, while R_4 is to limit the output voltage at the high end.

If the total resistance of the sampling circuit is R_T, then when the wiper of the pot is at the top of its travel, we have,

$$V_o \times \frac{R_T - R_3}{R_T} = 1.236 \text{ V} \quad \text{and hence} \quad \frac{R_T - R_3}{R_T} = \frac{1.236 \text{ V}}{2 \text{ V}} = 0.618$$

If I choose a 10 kΩ pot for RV_1, and ignore, for the moment, the resistance R_4 (which will, in any case, be small compared with RV_1 and R_3), then,

Regulated d.c. power supplies

$$\frac{10 \text{ k}\Omega}{10 \text{ k}\Omega + R_3} = 0.618 \text{ and } R_3 = \frac{10 \text{ k}\Omega - 6.18 \text{ k}\Omega}{0.618} = 6.2 \text{ k}\Omega$$

Any resistance less than this will allow the output voltage to fall to less than 2 V, so to ensure the full specified output voltage range, I choose the nearest preferred value below 6.2 kΩ, which is 5.6 kΩ.

At the low end of the pot (the high output voltage end),

$$\frac{R_4}{R_T} = \frac{1.236 \text{ V}}{20 \text{ V}} = 0.062, \text{ so } R_4 = 0.062 \times 15.6 \text{ k}\Omega = 970 \text{ }\Omega$$

The nearest preferred value is $R_4 = 1$ kΩ.

To check these choices,

$$V_o(\min) = V_{\text{ref}} \times \frac{R_T}{R_T - R_3} = 1.236 \text{ V} \times \frac{16.6 \text{ k}\Omega}{11 \text{ k}\Omega} = 1.86 \text{ V}$$

and

$$V_o(\max) = V_{\text{ref}} \times \frac{R_T}{R_4} = 1.236 \text{ V} \times \frac{16.6 \text{ k}\Omega}{1 \text{ k}\Omega} = 20.5 \text{ V}$$

These values therefore more than meet the specification for the output voltage range, and allow a margin of safety for component tolerances.

Overload protection

The overload protection sensing resistor R_S must be chosen to generate a voltage drop of 0.65 V at a load current of 2 A. Hence its value must be 0.325 Ω. When the current reaches 2 A, T3 starts to conduct and takes base current from T2, so preventing any further increase in load current.

Load regulation

Any change in output voltage with an increase of load current will be caused by (i) an increase in the base-emitter voltages of T1 and T2, (ii) an increase in the voltage drop across R_S and (iii) any change in the voltage reference.

These changes in the base emitter voltages and the change in the voltage across R_S will all be reduced by the feedback factor $(1 + A\beta)$ of the voltage stabilising circuit. The change in the reference voltage will not, since the purpose of the stabilising loop is to maintain the output of the sampling circuit equal to the voltage reference. In fact, any change in the reference voltage will be amplified by the ratio $V_{\text{out}}/V_{\text{ref}}$.

(i) The change in the base-emitter voltage of a transistor due to a change in the emitter current from I_1 to I_2 can be evaluated using the equation

$$\Delta V_{BE} = \frac{1}{K} \ln\left(\frac{I_2}{I_1}\right) \text{ (which was developed in the solution to Question 14.5).}$$

So, for an increase in load current from 50 mA to 1 A, the base-emitter voltage of T2 will increase by

$$\Delta V_{BE1} = \frac{1}{K}\ln\left(\frac{I_2}{I_1}\right) = \frac{1}{40\text{ V}^{-1}} \times \ln\left(\frac{1\text{ A}}{50\text{ mA}}\right) = 75\text{ mV}$$

and the base-emitter voltage of T2 will increase by

$$\Delta V_{BE2} = \frac{1}{40\text{ V}^{-1}} \times \ln\left(\frac{1\text{ A}/30}{50\text{ mA}/30}\right) = 75\text{ mV}.$$

So, the total voltage between the base of T1 and the emitter of T2 will increase by 150 mV when the load current increases from 50 mA to 1 A.

(ii) At the same time, the voltage drop across R_S will increase from 0.325 Ω × 50 mA (≈ 16 mV) to 0.325 Ω × 1 A = 325 mV a change of approximately 310 mV.

The total change in the voltage drop between the base of T1 and the output will therefore be 460 mV, but this will be reduced by the stabilising circuit which tries to compensate by changing the base voltage of T1. As a result, the change in output voltage due to these two causes will be reduced by the factor $(1 + A\beta)$ where A is the gain of the comparator amplifier multiplied by the gain of the Darlington pair, while β is the feedback factor, which in this case is the proportion of the output voltage fed back to the comparator amplifier.

(iii) There will be a change in the reference voltage when the output current changes from its minimum to its maximum value, due to the change in the average value of the voltage across C_1 as the ripple increases.

The average voltage across C_1 at 50 mA load current will be about 24 V, because the ripple will be very small. At 1 A, the ripple is 2 V p-p and so the average value is 23 V.

This 1 V change in the unstabilised voltage will cause a change in reference voltage due to the internal resistance of the reference source which is given by

$$\Delta V_{ref} = \Delta V_{C_1} \times \frac{r_{int}}{R_1 + R_2 + r_{int}} = 1\text{ V} \times \frac{5\text{ Ω}}{2\text{ kΩ} + 5\text{ Ω}} = 2.5\text{ mV}$$

When the output voltage is 20 V, this reference voltage change will be amplified, by the sampling and stabilising circuits, by the factor 20/1.236 = 16, and so the change in output voltage due to this cause will be 2.5 mV × 16 = 40 mV.

When the output voltage is 2 V, ΔV_{ref} will be amplified by the factor 2/1.236 = 1.6, and the change in output voltage due to this cause will be 2.5 mV × 1.6 = 4 mV.

Note that this is 0.2% of the output voltage in both cases, and would be the same irrespective of the actual output voltage setting.

The specification requires a load regulation of 0.5%, so the remaining 0.3% is available for the changes within the feedback loop.

When the output voltage is 2 V, this 0.3% represents 6 mV, so the 460 mV change inside the loop must be reduced by the factor $460 \div 6 = 77$, so $(1 + A\beta)$ must be at least 77. At this output voltage, $\beta = 1.236/2 = 0.62$ so, $(1 + 0.62A) \geq 77$ or $A \geq 123$.

When the output voltage is 20 V, this 0.3% represents 60 mV, so the 460 mV change inside the loop must be reduced by the factor $460 \div 60 = 7.7$. At this output voltage, $\beta = 1.236/20 = 0.062$ so, $(1 + 0.062A) \geq 7.7$ or $A \geq 108$, which is a less exacting requirement than at 2 V.

The load regulation requirement will therefore be met provided the gain of the comparator amplifier is at least 123.

Line regulation

If the supply voltage increases by 10%, the voltage at C_1 increases by 10% i.e. by 2.3 V. The bandgap reference voltage will then increase by $2.3 \text{ V} \times 5/2005 \approx 6$ mV. This change in the reference voltage will produce an output change of 6 mV $\times V_{out} \div 1.236$ V which represents a percentage change in output voltage of $6/1236 \times 100\% = 0.49\%$.

The line regulation requirement of the specification is therefore met.

[If the specification had not been met, consideration would have to be given to increasing the stability of the reference voltage source, for example by preceding the bandpass reference source with a Zener diode voltage reference. A two-stage reference of this sort will 'insulate' the final reference source from much of the change in the smoothed output of the bridge rectifier.]

The transistors

The overload protection circuit should limit the output transistor emitter current to 2 A, but for safety, an output transistor should be chosen which can carry a somewhat higher current than this. I suggest the following transistor parameters:

T2	maximum collector current	>2.5 A
	maximum collector-emitter voltage	>30 V
	maximum power dissipation	>48 W (with suitable heat sink)
		(see heat sink calculation)
T2	maximum collector current	>100 mA
	maximum collector voltage	>30 V
	maximum power dissipation	>3 W (with suitable heat sink)
T3	maximum collector current	>100 mA
	maximum collector voltage	>30 V
	maximum power dissipation	3 W (with suitable heat sink)

The transformer peak current

To calculate the charging time, again we represent the rectified half-wave by $v = V\sin\theta$, so that, for 2 V peak ripple, $\theta = \sin^{-1}(22/24) = 66°$ when the charging commences. Assuming charging ceases as soon as the half sine-wave starts to fall, charging ends when $\theta = 90°$.

Therefore, charging time = $\dfrac{90° - 66°}{180°} \times 10$ ms = 1.3 ms.

Capacitor C_1 discharges, in the worst case of normal operation, with a current of 1 A for a time of $(10 - 1.3)$ ms = 8.7 ms. The charge lost from the capacitor is therefore 8.7 mC, which is also the charge to be replaced during the 1.3 ms available for charging. The average charging current during this time is therefore 8.7 mC ÷ 1.3 ms = 6.7 A. The peak current is then a little less than twice this value, and therefore about 12 A.

[Under overload conditions, this current peak amplitude would, of course, be exceeded.]

The heat sink

The specification requires that the power unit should be able to work in ambient temperatures up to 40 °C. To prevent catastrophic failure of the output transistor, the required thermal resistance of the heat sink should be calculated under short-circuit conditions when the emitter current is 2 A and the collector-emitter voltage is 24 V. The power being dissipated is then 48 W. The allowable temperature rise is 120 °C − 40 °C = 80 °C. The required overall thermal resistance is therefore 80 °C ÷ 48 W = 1.7 °C W^{-1}, and the thermal resistance of the heat sink must be $(1.7 - 0.7)$ °C W^{-1} = 1.0 °C W^{-1}.

The power supply circuit, with component values added is shown in Figure 14.12

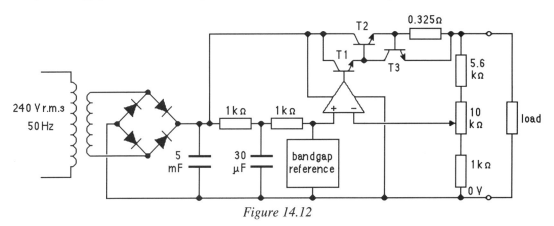

Figure 14.12

15 WAVEFORM GENERATION

QUESTIONS

15.1 (a) A quartz crystal has the equivalent circuit shown in Figure 15.1(a). What are the two resonant frequencies of the crystal and what is its Q factor?

(b) The crystal is used in the Pierce oscillator of Figure 15.1(b). What is the frequency of oscillation of the circuit?

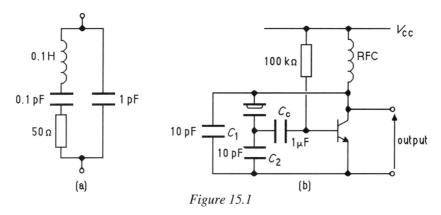

Figure 15.1

15.2 Calculate the frequency of oscillation of each of the square-wave oscillators shown in Figure 15.2

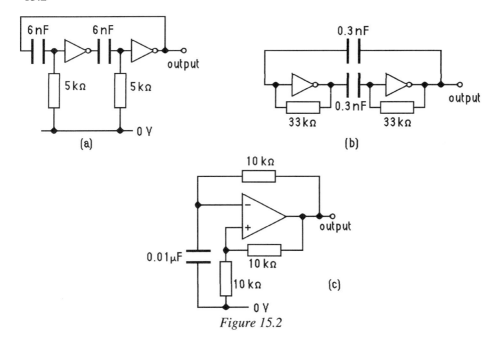

Figure 15.2

Waveform generation

15.3 (a) What are the waveforms at the two outputs of the circuit of Figure 15.3?

(b) What is the peak-to-peak amplitude of output 1?

(c) What is the peak-to-peak amplitude of output 2?

(d) What is the frequency of each output waveform?

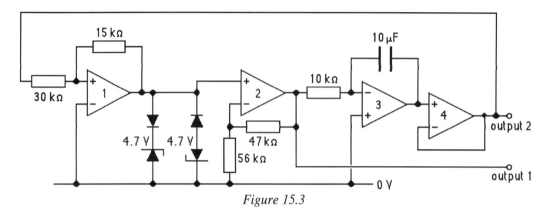

Figure 15.3

15.4 A Wien bridge oscillator is required to produce a low-distortion sinewave output, variable in frequency from 100 Hz to 10 kHz, and in amplitude from 0 to 5 V. The oscillator must operate from ±9 V supplies and must be able to provide the specified output amplitude with a load of 20 Ω. Sketch the required circuit and calculate the required resistor and capacitor values. Make any reasonable assumptions about the characteristics of other components which are necessary for the calculation of component values.

SOLUTIONS

15.1

(a)

The two resonant frequencies are :

(i) $$\omega_s = [LC_s]^{-\frac{1}{2}} = [0.1\text{ H} \times 0.1\text{ pF}]^{-\frac{1}{2}} = 10^7 \text{ rad s}^{-1}$$

and so $\quad f_s = \omega_s/2\pi = (10^7/2\pi) \text{ rad s}^{-1} = 1.59 \times 10^6 \text{ Hz}$

(ii) $$\omega_p = \left[\frac{LC_sC_p}{C_s + C_p}\right]^{-\frac{1}{2}} = \left[\frac{0.1\text{ H} \times 0.1\text{ pF} \times 1\text{ pF}}{0.1\text{ pF} + 1\text{ pF}}\right]^{-\frac{1}{2}} = 1.05 \times 10^7 \text{ rad s}^{-1}$$

and so $\quad f_p = \omega_p/2\pi = 1.05 \times 10^7/2\pi \text{ rad s}^{-1} = 1.67 \times 10^6 \text{ Hz}$

The Q factor of the crystal is:

$$Q = \omega_0 L/R \approx \left(10^7 \text{ rad s}^{-1} \times 0.1\text{ H}\right)/50\ \Omega = 2 \times 10^4$$

(b)

The frequency of oscillation of the circuit is:

$$\omega_R = \left[\frac{LC'C_s}{C' + C_s}\right]^{-\frac{1}{2}} \quad \text{where} \quad C' = \frac{C_1C_2}{C_1 + C_2} + C_p$$

So, $\quad C' = \dfrac{10\text{ pF} \times 10\text{ pF}}{20\text{ pF}} + 1\text{ pF} = 6\text{ pF}$

and $\quad \omega_R = \left[\dfrac{0.1\text{ H} \times 6\text{ pF} \times 0.1\text{ pF}}{6.1\text{ pF}}\right]^{-\frac{1}{2}} = 1.01 \times 10^7 \text{ rad s}^{-1}$

and $\quad f_R = \omega_R/2\pi = 1.01 \times 10^7/2\pi \text{ rad s}^{-1} = 1.60 \times 10^6 \text{ Hz}$

Question 15.2

(a)

The period of the waveform $T \approx 2CR = 2 \times 6\text{ nF} \times 5\text{ k}\Omega = 60\ \mu\text{s}$,

and so, $\quad f = 1/T \approx 17 \text{ kHz}$

(b)

The period of the waveform $T \approx CR = 0.3\text{ nF} \times 33\text{ k}\Omega = 10\ \mu\text{s}$,

and so $\quad f = 1/T \approx 100 \text{ kHz}$

Waveform generation

(c)

The period of the waveform $T = 2CR \ln\left[\dfrac{1+k}{1-k}\right]$ where $k = R_1/(R_1 + R_F) = \tfrac{1}{2}$

so, $T = 2 \times 0.01\,\mu\text{F} \times 10\,\text{k}\Omega \times \ln\left[\dfrac{3/2}{1/2}\right] = 2 \times 10^{-4} \ln 3 = 2.2 \times 10^{-4}\,\text{s}$

and $f = 1/T \approx 4.5\,\text{kHz}$.

Question 15.3

(a)

Output 1 is a square wave and output 2 is a triangular wave.

(b)

The square-wave at the output of op-amp 1 is limited by the combinations of Zener diode plus diode to about $\pm(4.7 + 0.6)$ V i.e. to ± 5.3 V.

Output 1 is the output of op-amp 2, which is a non-inverting amplifier with a gain of $(47 + 56) \div 56 = 1.84$. Output 1 is therefore a square-wave with peak values $\pm(5.3\,\text{V} \times 1.84) = \pm 9.75\,\text{V}$.

The peak-to-peak amplitude of output 1 is therefore 19.5 V.

(c)

The output of op-amp 3 will be the voltage across the feedback capacitor of the integrating amplifier.

Because the inverting input of the op-amp is a virtual earth, if I is the current flowing through the input resistor R and through the feedback capacitor C, then.

$$I = V_{in}/R \text{ and } I = \frac{dQ}{dt} = \frac{d}{dt}(-CV_{out}) \text{ or } V_{out} = -\frac{1}{C}\int I\,dt$$

so, $V_{out} = -\dfrac{1}{C}\int \dfrac{V_{in}}{R}\,dt = -\dfrac{1}{CR}\int V_{in}\,dt$

(which is why the circuit is called an integrator).

In this circuit, V_{in} is output 1, and is therefore either $+9.75$ V or -9.75 V.

When $V_{in} = +9.75$ V, $V_{out} = -\dfrac{1}{CR}\int 9.75\,\text{V}\,dt = -\dfrac{9.75\,\text{V}}{10\,\mu\text{F} \times 10\,\text{k}\Omega} \times t = -97.5\,\text{V} \times t$

and the output of op-amp 3 ramps *down* at 97.5 V s^{-1}.

Similarly, when the input voltage is -9.75 V, the output ramps *up* at 97.5 V s^{-1}.

When the input to op-amp 3 is positive, the output of op-amp 1 is also positive, and remains so until its non-inverting input terminal reaches zero volts, when the amplifier output changes sign.

The input will become zero when $V_{in}R_F = -V_{out}R_{in}$, i.e. when $V_{in} = -(V_{out}R_{in})/R_F$.

For the component values in the circuit, the output changes sign when

$$V_{in} = -\frac{5.3 \text{ V} \times 30 \text{ k}\Omega}{15 \text{ k}\Omega} = -10.6 \text{ V}.$$

Thus output 2 will change slope at +10.6 V and −10.6 V, so that the peak-to-peak amplitude of the triangle wave output is 21.2 V.

(d)

Since the total voltage change of the triangle wave output is 21.2 volts, and since the rate of change of voltage is 97.5 V s^{-1}, the time taken for one complete cycle is 21.2 V ÷ 97.5 V s^{-1} × 2 = 0.435 s.

The frequency of oscillation is therefore 2.3 Hz.

Question 15.4

The output current amplitude requirement of 5 V ÷ 20 Ω = 250 mA is larger than can be obtained from a typical op-amp, so either a special purpose audio frequency power amplifier is required, or an op-amp feeding a power output stage of the type described in Chapter 13.

A feasible block diagram of the oscillator is shown in Figure 15.4

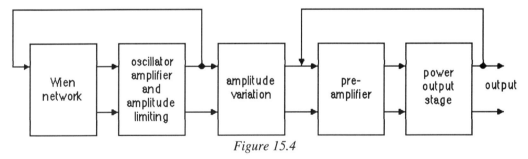

Figure 15.4

The Wien network

The Wien network will need variable components to allow frequency variation. Normally, capacitors are switched to establish decade ranges, while pots are used as variable resistors to give continuous frequency control within each decade.

A circuit for the Wien network is shown in Figure 15.5

The required frequency range is from 100 Hz to 10 kHz, so there will be two decade ranges selected by SW1, with $C_2 = C_4 = 10C_1 = 10C_3$. The pots RV1(a) and RV1(b) are mechanically "ganged" and must be capable of changing the total resistance in each leg by a factor of 10. Since the

Waveform generation

resisances R_1 and R_2 are equal, the maximum resistance of RV1(a) and RV1(b) must be at least 9 times R_1.

If we make the pot resistance 10 times R_1, then there will be a small overlap of the two decade ranges, which can make operation of the oscillator more convenient.

Convenient values to use are 10 kΩ for R_1 and R_2, and 100 kΩ pots for RV1(a) and RV1(b).

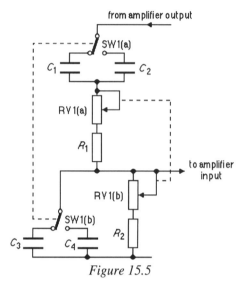

Figure 15.5

Since, for the Wien bridge oscillator, $f = 1/(2\pi CR)$, the frequency will be at its highest when C and R are at their minimum values.

With the variable resistors set to zero, the required value of C can be found from the equation $C = 1/2\pi Rf$.

For a frequency of 10 kHz,

$$C_1 = C_3 = \frac{1}{2\pi \times 10 \text{ k}\Omega \times 10^4 \text{ Hz}} = 1.6 \text{ nF}$$

For a frequency of 1 kHz,

$$C_1 = C_3 = \frac{1}{2\pi \times 10 \text{ k}\Omega \times 10^3 \text{ Hz}} = 16 \text{ nF}$$

When the variable resistors are set at their maximum value, $R = 110$ kΩ and so, on the higher frequency range,

$$f = \frac{1}{2\pi \times 1.6 \text{ nF} \times 110 \text{ k}\Omega} = 909 \text{ Hz}$$

and on the lower range,

$$f = \frac{1}{2\pi \times 16 \text{ nF} \times 110 \text{ k}\Omega} = 90.9 \text{ Hz}$$

The oscillator amplifier

The oscillator amplifier with its amplitude limiting circuit is shown in Figure 15.6.

The gain at low output amplitude is required to be about 3.2 to ensure that oscillation starts reliably, and the gain should fall as the amplitude increases until it falls just below 3 before limitations caused by power supply voltages occur.

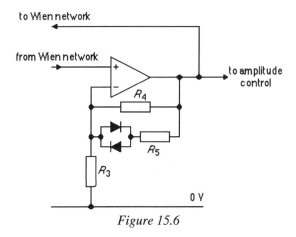

Figure 15.6

For the amplifier to have a gain of 3.2 at low signal amplitude, i.e. before the diodes start conducting to any significant extent, $(R_3 + R_4) \div R_3$ must equal 3.2, which means that $R_4 \div R_3 = 2.2$.

We can make $R_3 = 10$ kΩ and $R_4 = 22$ kΩ. These values will not place any significant load on the output of the op-amp, and yet are not so large as to be a source of significant noise pick-up.

To ensure that the gain falls below 3 at higher signal amplitudes, if the total feedback resistance is R_F, then $(R_F + R_3)/R_3$ must be less than 3, or R_F/R_3 must be less than 2.

Since $R_3 = 10$ kΩ, R_F must be less than 20 kΩ when the diodes are conducting. Then,

$$\frac{R_4 R_5}{R_4 + R_5} < 20 \text{ k}\Omega \text{ and since } R_4 = 22 \text{ k}\Omega, \ R_5 < 220 \text{ k}\Omega.$$

So we can make $R_5 = 200$ kΩ, and the gain will drop to 3 when the diode looks like a resistance of 20 kΩ, i.e. when $I_D = 0.6$ V $\div$ 20 kΩ = 30 µA.

When the current in the 200 kΩ resistor is 30 µA, the voltage across it will be 6 V, so adding the diode forward voltage of about 0.6 V, the output amplitude will be 6.6 V.

The output of the oscillator amplifier will therefore limit at an amplitude of about 6.6 V.

The amplitude control and output amplifier

A possible configuration for the amplitude control and output amplifier of the oscillator is shown in Figure 15.7. Other configurations are, of course, possible, as you saw in Chapter 13.

Waveform generation

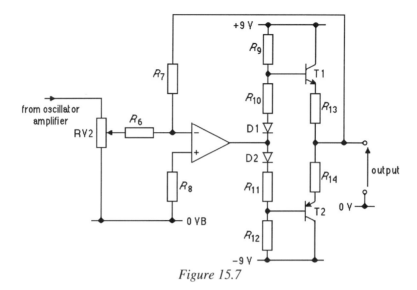

Figure 15.7

The amplitude control pot RV2 must not unduly load the oscillator amplifier, and must not be unduly loaded by the input resistor of the pre-amplifier, so a reasonable value is 10 kΩ. Resistor R_6 can then be made 100 kΩ without the load on RV2 being significant.

Since the amplitude of the sinewave from the oscillator amplifier is about 6.6 V, the gain of the output amplifier could be made less than 1, but since the amplitude calculation is only approximate, it is probably safer to give the output amplifier a gain of 1 by making R_7 equal to R_6 at 100 kΩ.

R_8 will need to be equal to the parallel combination of R_6 and R_7 so as to minimise the effect of input bias currents, so R_8 will be 50 kΩ.

I will assume that T1 and T2 have a current gain β of 50 and a V_{BE} of 0.7 V when conducting, and that D1 and D2 have a forward voltage drop V_D of 0.6 V.

To reduce the possibility of distortion, I will assume Class A operation of the double emitter-follower output stage.

Since the peak load current is 250 mA, the required quiescent current is 125 mA.

(The remainder of this calculation follows the steps of the solution to Question 13.1.)

To minimise distortion due to the variation in output resistance of the emitter followers, R_{13} and R_{14} should each be $0.6 r_{eo}$, where r_{eo} is the emitter follower output resistance in the quiescent state.

Since the op-amp output is fairly low resistance, it approximates to an ideal signal source and the emitter follower output resistance is then

$$r_{eo} = 1/g_m = 1/KI_c = 1/\left(40 \text{ V}^{-1} \times 125 \text{ mA}\right) = 0.2 \text{ } \Omega$$

and

$$R_{13} = R_{14} = 0.6 \times 0.2 \text{ } \Omega = 0.12 \text{ } \Omega.$$

The required values of R_9 and R_{12} can be found from the peak signal conditions.

At a 5 V positive excursion of the input signal, the T1 emitter current = 250 mA, and so the peak value of the base current is 5 mA. At the same time, the base of T1 will be at a voltage equal to the load voltage plus the base-emitter voltage of T1. (The voltage drop across R_{13} is negligible in comparison with the other voltages.) The base voltage of T1 is therefore 5.7 V, leaving a voltage across R_9 of 9 V − 5.7 V = 3.3 V. The maximum value of R_9 is therefore 3.3 V ÷ 5 mA = 0.66 kΩ.

I will choose $R_9 = R_{12} = 620$ Ω, which is the nearest preferred value below 660 Ω.

To find the value of R_{10} and R_{11} we must consider the quiescent condition.

In the quiescent condition, the base current of T1 = 125 mA ÷ 50 = 2.5 mA, and the base voltage is 0.7 V (since the output voltage is zero). If I_D is the current through the diodes in the quiescent condition, then

$$(I_D + 2.5 \text{ mA}) \times 620 \text{ }\Omega = 9 \text{ V} - 0.7 \text{ V}$$

$$I_D + 2.5 \text{ mA} = 8.3 \text{ V}/620 \text{ }\Omega = 13.4 \text{ mA}$$

$$I_D = 10.9 \text{ mA}$$

Hence, $\quad R_{10} = R_{11} = (0.7 \text{ V} - 0.6 \text{ V})/10.9 \text{ mA} = 9 \text{ }\Omega$

This completes the specification of the resistor and capacitor values.

16 THE HIGH-FREQUENCY BEHAVIOUR OF TRANSISTORS

QUESTIONS

16.1 A transistor has the following parameters:

$C_C = 8$ pF at zero bias $\phi_{(CB)} = -0.65$ V

$C_{et} = 5$ pF at zero bias $\phi_{(EB)} = -0.8$ V

$\tau_t = 1$ ns $\beta = 200$

$V_A = 80$ V $r_b = 50\ \Omega$.

Calculate the values of C_C and C_i when $V_{CE} = 15$ V, $I_C = 1$ mA and $V_{BE} = 0.65$ V.

16.2 (a) The angular frequency ω_1 corresponding to the unity-current-gain frequency f_1 of a transistor is quoted on page 617 of the textbook (equation 16.13) as

$$\omega_1(C_i + C_c) = |g_m - j\omega_1 C_c|.$$

Use this equation to show that

$$f_1 = \frac{g_m}{2\pi\sqrt{C_i(C_i + 2C_c)}}.$$

(b) Sketch the straight-line approximation to the Bode gain plot of h_{fe} for the transistor and operating point of Question 16.1, specifying the low-frequency magnitude of h_{fe}, the β cut-off frequency f_β, and the transition frequency f_T.

(c) Calculate the range of frequencies termed "intermediate frequencies" on page 615 of the textbook.

16.3 The transistor of Question 16.1 is placed in the circuit of Figure 16.1. The input voltage source has negligible internal resistance. Using the results of the analysis in Appendix 16A of the textbook, find

(a) the voltage gain of the amplifier at a frequency of 10 kHz;
(b) the low-frequency 3 dB point of the amplifier;
(c) the high-frequency 3 dB point of the amplifier;
(d) the gain-bandwidth product.

The high-frequency behaviour of transistors

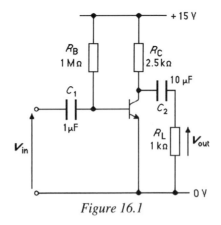

Figure 16.1

16.4 The amplifier of Figure 16.1 is now connected to a signal source having an internal resistance of 1 kΩ. Re-calculate the voltage gain, the two 3 dB points and the gain-bandwidth product.

16.5 An operational amplifier has the circuit of Figure 16.2. Each transistor can be assumed to have the following parameters:

$\beta = 100$, $VA = 100$ V, $C_C = 5$ pF, $f_T = 120$ MHz, $r_b = 100$ Ω.

The op-amp can be assumed to have zero output voltage when V_{in} is zero.

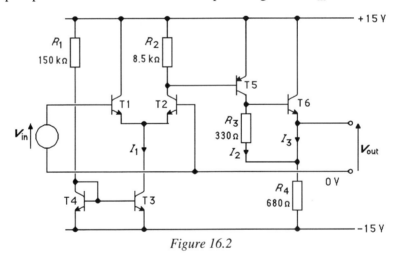

Figure 16.2

(a) Calculate the low-frequency open-circuit gain of the long-tailed pair (i.e. the gain with no load other than R_2), assuming that the signal source has negligible internal resistance.

(b) Represent the long-tailed pair by its low-frequency Thévenin equivalent circuit as shown in Figure 16.3 (where A is the low-frequency gain calculated in (a) and g_{out} is the output conductance of the long-tailed pair) and hence draw the hybrid-π small-signal equivalent circuit of the common-emitter amplifier (T5) connected to the long-tailed pair. The equivalent circuit should be in a form which will allow the results of the analysis of Appendix 16A to be used to calculate circuit performance.

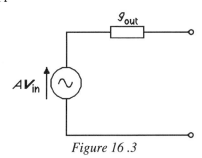

Figure 16.3

(c) Assuming that the input conductance of the emitter follower (including R_3) is 22 μS (*cf* the solution of Question 10.10 part (b)), calculate (i) the overall low-frequency gain of the amplifier and (ii) the 3 dB frequency of the Bode gain plot.

(d) A capacitor is to be added to the amplifier, connected between the base and collector of T5, to reduce the 3 dB frequency to 150 Hz. What value capacitor is required?

The high-frequency behaviour of transistors

SOLUTIONS

Question 16.1

When the bias is zero, using the equation

$$C_C = \text{const} \times (-\phi - V_{BC})^{-\frac{1}{3}}$$

$$8 \text{ pF} = \text{const} \times (0.65)^{-\frac{1}{3}}$$

and so $\quad \text{const} = 8 \text{ pF} \times (0.65)^{\frac{1}{3}}$

The reverse bias on the collector-base junction = 15 V − 0.65 V = 14.35 V, so

$$C_C = 8 \text{ pF} \times (0.65)^{\frac{1}{3}} \times (0.65 + 14.35)^{-\frac{1}{3}} = 2.8 \text{ pF}$$

Now $\quad C_i = C_{et} + C_{de}$

where $\quad C_{de} = g_m \tau_t = K I_C \tau_t$

and $\quad C_{et} = \text{const} \times (-\phi - V_{BE})^{-\frac{1}{3}}$

With zero bias, $5 \text{ pF} = \text{const} \times (0.8)^{-\frac{1}{3}}$ and so $\text{const} = 5 \text{ pF} \times (0.8)^{\frac{1}{3}}$

With 0.65 V forward bias,

$$C_{et} = 5 \text{ pF} \times (0.8)^{\frac{1}{3}} \times (0.8 - 0.65)^{-\frac{1}{3}} = 8.7 \text{ pF}$$

$$C_{de} = K I_C \tau_t = 40 \text{ V}^{-1} \times 1 \text{ mA} \times 1 \text{ ns} = 40 \text{ pF}$$

and so, $\quad C_i = C_{et} + C_{de} = 8.7 \text{ pF} + 40 \text{ pF} \approx 49 \text{ pF}.$

Question 16.2

(a)

Using the equation $\quad \omega_1 (C_i + C_C) = |g_m - j\omega_1 C_C|$

and substituting $\quad |g_m - j\omega_1 C_C| = \sqrt{g_m^2 + \omega_1^2 C_C^2}$

we obtain

$$\omega_1^2 (C_i + C_C)^2 = g_m^2 + \omega_1^2 C_C^2$$

$$\omega_1^2 (C_i^2 + 2 C_i C_C + C_C^2 - C_C^2) = g_m^2$$

hence $\quad \omega_1 = \dfrac{g_m}{\sqrt{C_i^2 + 2 C_i C_C}}$

and
$$f_1 = \frac{g_m}{2\pi\sqrt{C_i(C_i + 2C_C)}}$$

(b)

At low frequencies, $|h_{fe}| = h_{fe} = \beta = 200$ and so $20\log_{10}|h_{fe}| = 46$ dB.

At intermediate frequencies, where $\omega C_C \ll g_m$ and $g_i \ll \omega(C_i + C_C)$,

the equation
$$h_{fe} = \frac{-g_m + j\omega C_C}{g_i + j\omega(C_i + C_C)}$$

reduces to
$$|h_{fe}| = \frac{g_m}{\omega(C_i + C_C)} = \frac{40 \text{ mA V}^{-1}}{\omega \times (49 + 2.8) \times 10^{-12} \text{ F}} = \frac{0.77 \times 10^9}{\omega}$$

This, on the Bode plot of $|h_{fe}|$, will be a straight line at a slope of -20 dB per decade. It will meet the horizontal line representing the low-frequency gain at the frequency given by $200 = \dfrac{0.77 \times 10^9}{\omega}$ i.e.

at the frequency $f_\beta = \omega/2\pi = \dfrac{0.77 \times 10^9}{2\pi \times 200}$ Hz $= 0.61$ MHz.

The transition frequency f_T is the frequency at which the -20 dB per decade slope would meet the 0 dB (i.e. unity gain) line if it continued at that slope. Hence,

$$1 = \frac{0.77 \times 10^9}{\omega_T} \quad \text{and} \quad f_T = \omega_T/2\pi = \frac{0.77 \times 10^9}{2\pi} = 122.5 \text{ MHz}.$$

The unity-gain frequency is given by the equation developed in part (a), i.e.

$$f_1 = \frac{g_m}{2\pi\sqrt{C_i(C_i + 2C_C)}} = \frac{40 \text{ mA V}^{-1}}{2\pi\sqrt{49 \text{ pF}(49 \text{ pF} + 5.6 \text{ pF})}} = 123 \text{ MHz}$$

The sketch of the straight-line approximation to $20\log_{10}|h_{fe}|$ is shown in Figure 16.4.

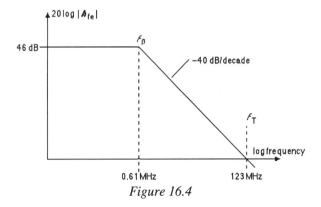

Figure 16.4

The high-frequency behaviour of transistors

(c)

The "intermediate frequency" range is defined as that range of frequencies for which $\omega C_C \ll g_m$ and $g_i \ll \omega(C_i + C_C)$.

If $\omega C_C \ll g_m$ then $\omega \ll (40 \text{ mA V}^{-1})/(2.8 \times 10^{-12} \text{ F}) = 14 \times 10^9 \text{ rad s}^{-1}$ and $f \ll 2.3$ GHz.

If $g_i \ll \omega(C_i + C_C)$ then since $g_i = g_m/\beta$,

$$\omega \gg (40 \text{ mA V}^{-1})/[200 \times (49 \text{ pF} + 2.8 \text{ pF})] = 3.9 \times 10^6 \text{ rad s}^{-1} \text{ and } f \gg 0.6 \text{ MHz}.$$

The intermediate frequency range for which the approximate equation for $|h_{fe}|$ is valid is:

$$0.6 \text{ MHz} \gg f \gg 2.3 \text{ GHz}.$$

Question 16.3

(a)

At 10 kHz, the reactance of $C_1 = X_{C_1} = 1/(2\pi f C_1) = 1/(2\pi \times 10^4 \text{ Hz} \times 10^{-6} \text{ F}) = 16 \, \Omega$

At the same frequency, the reactance of $C_2 = X_{C_2} = 1/(2\pi \times 10^4 \text{ Hz} \times 10^{-5} \text{ F}) = 1.6 \, \Omega$

Compared with other resistance values in the circuit, the reactances of capacitors C_1 and C_2 can be neglected in the calculations.

The d.c. base current $I_B = (15 \text{ V} - 0.65 \text{ V}) \div 1 \text{ M}\Omega \approx 14.4 \text{ μA}$, so the d.c. collector current $I_C = 14.4 \text{ μA} \times 200 = 2.8 \text{ mA}$ and the collector-emitter voltage $V_{CE} = 15 \text{ V} - (2.8 \text{ mA} \times 2.5 \text{ k}\Omega) = 8 \text{ V}$.

The transistor parameters are therefore:

$$g_m = KI_C = 40 \text{ V}^{-1} \times 2.8 \text{ mA} = 112 \text{ mA V}^{-1}$$

$$g_i = g_m/\beta = 112 \text{ mA V}^{-1} \div 200 = 0.56 \text{ mS} \quad (r_i \approx 1.8 \text{ k}\Omega)$$

$$g_o = I_C/(V_A + V_{CE}) = 3 \text{ mA} \div (80 \text{ V} + 8 \text{ V}) = 34 \text{ μS} \quad (r_o = 29 \text{ k}\Omega).$$

$$C_C = 8 \text{ pF} \times (0.65 \text{ V})^{\frac{1}{3}} \times (0.65 \text{ V} + (8 - 0.65) \text{ V})^{-\frac{1}{3}} = 3.5 \text{ pF}$$

$$C_i = C_{et} + C_{de} = 8.7 \text{ pF} + 112 \text{ mA V}^{-1} \times 1 \text{ ns} = 121 \text{ pF}$$

Figure 16.5 is the small signal equivalent circuit of the amplifier, neglecting C_1 and C_2. Comparing this circuit with that of Figure 16.10(b) of the textbook (page 622), you can see that they are the same except for the addition in Figure 16.5 of R_B and R_L.

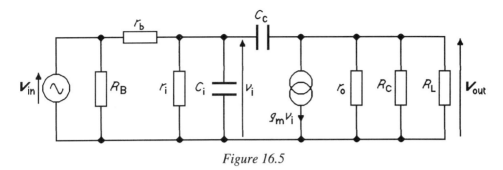

Figure 16.5

Figure 16.6 is an alternative equivalent circuit to that of Figure 16.5 provided $g_{C(tot)} = g_o + g_C + g_L$, where $g_C = 1/R_C$ and $g_L = 1/R_L$. The presence of R_B does not affect the analysis, because it is across the zero source resistance.

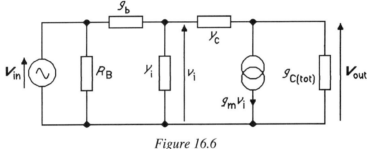

Figure 16.6

For Figure 16.6 to represent the circuit of Figure 16.1, the following values must be used

$$g_C = 1/R_C = 400 \text{ μS}, \quad g_L = 1/R_L = 1000 \text{ μS}, \text{ and } g_{C(tot)} = g_o + g_L + g_C = 1434 \text{ μS}$$

If the effect of the transistor internal capacitances can be neglected at 10 kHz (we will check later), then the gain at this frequency will be:

$$A_V = \frac{-g_m}{g_{C(tot)} \times (1 + r_b g_i)} = \frac{-112 \text{ mA V}^{-1}}{1434 \text{ μS} \times (1 + 50 \text{ Ω} \times 560 \text{ μS})} = -76$$

(b)

The low-frequency fall-off in gain will be due to the reactances of capacitors C_1 and C_2. To calculate the low-frequency 3 dB point, we need to find the time constants of the circuits associated with C_1 and C_2.

The input time constant will be $C_1(r_b + r_i) = 1 \text{ μF} \times (50 \text{ Ω} + 1.8 \text{ kΩ}) = 1.85 \text{ ms}$.

The output time constant will be $C_2/g_{C(tot)} = 10 \text{ μF}/1434 \text{ μS} = 7 \text{ ms}$.

Clearly, the smaller time constant will correspond to the higher corner frequency, which is:

The high-frequency behaviour of transistors

$$f = \omega/2\pi = 1/2\pi T = 1/(2\pi \times 1.85 \text{ ms}) = 86 \text{ Hz}$$

The second corner frequency (due to the 7 ms time constant) will occur at 23 Hz and will affect the gain slightly at 86 Hz, perhaps lowering the gain by 0.5 dB and hence raising the 3 dB frequency to around 100 Hz.

(c)

The high-frequency 3 dB point can be found using equations 16A.10, 16A.12 and 16A.13 of Appendix 16A of the textbook.

Equation 16A.13 is,

$$P = r_b \left(C_i + \frac{(g_i + g_m)C_C}{g_{C(tot)}} \right) + \frac{C_C}{g_{C(tot)}}$$

$$= 50 \, \Omega \left(121 \text{ pF} + \frac{(0.56 \text{ mS} + 112 \text{ mA V}^{-1}) \times 3.5 \text{ pF}}{1434 \text{ μS}} \right) + \frac{3.5 \text{ pF}}{1434 \text{ μS}}$$

$$= 50 \, \Omega \times (121 + 275) \text{ pF} + 2440 \text{ ps}$$

$$= 22\,240 \text{ ps}$$

Combining equations 16A.10 and 16A.12.

$$\omega_1 = \frac{1 + r_b g_i}{P} = \frac{1 + 50 \, \Omega \times 0.56 \text{ mS}}{22240 \text{ ps}} = 4.6 \times 10^7 \text{ rad s}^{-1}$$

and so $\quad f_1 = \omega_1/2\pi = 7.3 \text{ MHz}.$

(The 10 kHz frequency at which the gain was calculated is almost 3 decades below the high-frequency 3 dB point and two decades above the low-frequency 3 dB point, and is therefore well within the flat portion of the amplifier's frequency response. The assumption made previously that the effect of the transistor's internal capacitances could be ignored at 10 kHz was clearly valid.)

(d)

The bandwidth of the amplifier = 7.3 MHz – 100 Hz = 7.3 MHz.

The amplitude of the amplifier gain in the flat portion of the response = 76.

Therefore the gain-bandwidth product = 555 MHz.

Question 16.4

If the amplifier of Figure 16.1 is now fed from a signal source having a source resistance of 1 kΩ, the equivalent circuit of Figure 16.5 needs modifying to that of Figure 16.7, where R_S is the resistance of the signal source.

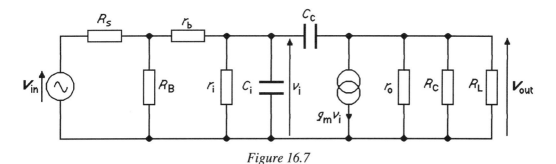

Figure 16.7

Since $R_B = 1$ MΩ and $R_s = 1$ kΩ, the attenuation of that network will be only 0.999, while the parallel combination of R_s and R_B will have a resistance of 0.999 kΩ. The effect of R_B can therefore be ignored, and the presence of R_s then simply increases the effective value of r_b by 1 kΩ.

The calculations of Question 16.3 can therefore be repeated, with a value of r_b of 1.05 kΩ.

The gain in the flat portion of the amplifier response will be:

$$A_v = \frac{-g_m}{g_{C(tot)} \times (1 + r_b g_i)} = \frac{-112 \text{ mA V}^{-1}}{1434 \text{ }\mu\text{S} \times (1 + 1050 \text{ }\Omega \times 560 \text{ }\mu\text{S})} = -49$$

The time constant of the input circuit will be:

$$C_1(r_b + r_i) = 1 \text{ }\mu\text{F} \times (1050 \text{ }\Omega + 1.8 \text{ k}\Omega) = 2.85 \text{ ms}$$

which gives a corner frequency of $f = 1/(2\pi \times 2.85 \text{ ms}) = 56 \text{ Hz}$.

The corner frequency of the output circuit remains 23 Hz.

At 56 Hz, the 23 Hz corner frequency will reduce the gain by about 1 dB, so making the actual 3 dB frequency about 1.4×56 Hz = 78 Hz, so the low frequency 3 dB point is not significantly affected by the increased value of r_b.

The high-frequency 3 dB point is found as in Question 16.3 part (c).

$$P = r_b\left(C_i + \frac{(g_i + g_m)C_c}{g_{C(tot)}}\right) + \frac{C_c}{g_{C(tot)}}$$

$$= 1050 \text{ }\Omega\left(121 \text{ pF} + \frac{(0.56 \text{ mS} + 112 \text{ mA V}^{-1}) \times 3.5 \text{ pF}}{1434 \text{ }\mu\text{S}}\right) + \frac{3.5 \text{ pF}}{1434 \text{ }\mu\text{S}}$$

$$= 1050 \text{ }\Omega \times (121 + 275) \text{ pF} + 2440 \text{ ps}$$

$$= 418 \text{ ns}$$

The high-frequency behaviour of transistors

$$\omega_1 = \frac{1+r_b g_i}{P} = \frac{1+1050\ \Omega \times 0.56\ \text{mS}}{418\ \text{ns}} = 3.8 \times 10^6\ \text{rad s}^{-1}$$

$$f_1 = \omega_1/2\pi = 0.6\ \text{MHz}.$$

The gain-bandwidth product is 0.6 MHz × 49 ≈ 30 MHz.

The presence of a significant signal source resistance not only reduces the gain of the amplifier, it also substantially reduces the bandwidth because of the interaction of the source resistance with the internal capacitances of the transistor.

Question 16.5

In order to calculate the relevant transistor parameters from the information given, we first need to calculate the d.c operating points of transistors T2 and T5.

$$(I_3 + I_2) \times 680\ \Omega = 15\ \text{V}, \text{ so } (I_3 + I_2) = 22\ \text{mA}$$

$$I_2 \times 330\ \Omega = 0.65\ \text{V}, \text{ so } I_2 \approx 2\ \text{mA and } I_3 = 20\ \text{mA}.$$

$$I_{C5} = I_2 + I_{B6} = 2\ \text{mA} + 20\ \text{mA}/100 = 2.2\ \text{mA}$$

$$I_{B5} = 2.2\ \text{mA}/100 = 22\ \mu\text{A}$$

$$I_1 = \text{current in } R_1 = (30\ \text{V} - 0.65\ \text{V}) \div 150\ \text{k}\Omega = 196\ \mu\text{A}$$

$$I_{R_2} = I_1/2 - I_{B5} = (196/2)\ \mu\text{A} - 22\ \mu\text{A} = 76\ \mu\text{A}$$

$$V_{BE5} = V_{R_2} = 76\ \mu\text{A} \times 8.5\ \text{k}\Omega = 0.65\ \text{V}$$

(a)

For transistor T2, $g_{m2} = KI_{C2} = 40\ \text{V}^{-1} \times 98\ \mu\text{A} = 3.9\ \text{mA V}^{-1}$

$$g_{o2} = I_{C2}/(VA + V_{CE2}) = (98\ \mu\text{A})/(100\ \text{V} + 15\ \text{V}) = 0.85\ \mu\text{S}.$$

$$g_{i2} = g_{m2}/\beta = (3.9\ \text{mA V}^{-1})/100 = 39\ \mu\text{S}$$

With no load other than R_2, the low-frequency voltage gain of the double emitter-follower is

$$A = \tfrac{1}{2} \times \frac{g_{m2}}{g_{o2}+g_2} \times \frac{1}{1+r_b g_i} = \tfrac{1}{2} \times \frac{3.9\ \text{mA V}^{-1}}{(0.85+118)\ \mu\text{S}} \times \frac{1}{1+100\ \Omega \times 39\ \mu\text{S}} = 16.3$$

The output conductance of the double-emitter follower is the sum of g_2 and g_{o2}, i.e. 119 μS.

(b)

The full small-signal equivalent circuit will be as shown in Figure 16.8.

The results of Appendix 16A can now be applied to this circuit, replacing r_b with (8.4 kΩ + r_b) and replacing $g_{C(tot)}$ with ($g_{o5} + g_{in6}$).

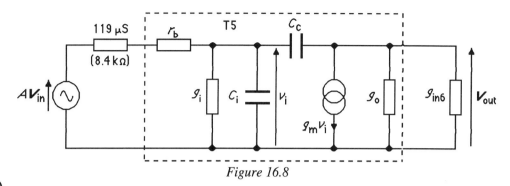

Figure 16.8

(c)

For transistor T5, from the transistor data given,

$$g_{m5} = KI_{C5} = 40 \text{ V}^{-1} \times 2.2 \text{ mA} = 88 \text{ mA V}^{-1}$$

$$g_{o5} = I_{C5}/(V_A + V_{CE5}) = (2.2 \text{ mA})/(100 \text{ V} + (15 - 0.65) \text{ V}) = 19 \text{ μS}.$$

$$g_{i5} = g_{m5}/\beta = (88 \text{ mA V}^{-1})/100 = 880 \text{ μS}$$

Using equation 16.11 of the textbook (page 616),

$$C_i + C_C = \frac{g_{m5}}{2\pi f_T} = \frac{88 \text{ mA V}^{-1}}{2\pi \times 120 \text{ MHz}} = 117 \text{ pF},$$

and since $C_C = 5$ pF, $C_i = 112$ pF.

(i) The low-frequency gain of the combination of the two stages is

$$\frac{V_{out}}{V_{in}} = A \times \frac{-g_m}{g_{o5} + g_{in6}} \times \frac{1}{1 + (r_{b5} + 8.4 \text{ kΩ}) \times g_{i5}}$$

$$= -16.3 \times \frac{88 \text{ mA V}^{-1}}{19 \text{ μS} + 22 \text{ μS}} \times \frac{1}{1 + (100 \text{ Ω} + 8.4 \text{ kΩ}) \times 880 \text{ μS}}$$

$$= -16.3 \times 2.15 \times 10^3 \times 0.118$$

$$\approx -4100$$

Since the gain of the emitter follower stage is 1, the overall low-frequency gain of the amplifier is also -4100.

(ii) Using equation 16A.13,

$$P = r_b \left(C_i + \frac{(g_i + g_m)C_C}{g_{C(tot)}} \right) + \frac{C_C}{g_{C(tot)}}$$

The high-frequency behaviour of transistors

$$= 8.5 \text{ k}\Omega \left(112 \text{ pF} + \frac{\left(0.88 \text{ mS} + 88 \text{ mA V}^{-1}\right) \times 5 \text{ pF}}{(19+22) \text{ μS}} \right) + \frac{5 \text{ pF}}{(19+22) \text{ μS}}$$

$$= 8.5 \text{ k}\Omega \times (112 + 10840) \text{ pF} + 122 \text{ ns}$$

$$= 93.2 \text{ μs}$$

$$\omega_1 = \frac{1 + r_b g_i}{P} = \frac{1 + 8.5 \text{ k}\Omega \times 0.88 \text{ mS}}{93.2 \text{ μs}} = 91 \times 10^3 \text{ rad s}^{-1}$$

$$f_1 = \omega_1 / 2\pi = 14.5 \text{ kHz}.$$

(d)

To reduce the high-frequency 3dB point to 150 Hz, P must be increased by a factor of $14.5 \times 10^3 \div 150 \approx 97$.

Since $r_{b5} \ll r_{out(2)}$, and since C_C is the major contributor to the value of P, increasing C_C by a factor of 97 will increase P by about 97.

So we need to connect a capacitor of $97 \times 5 \text{ pF} = 490 \text{ pF}$ between the base and collector of T5.

17 INTERCONNECTIONS

QUESTIONS

17.1 A signal source has an output voltage of 30 mV peak, and an output resistance of 10 kΩ. It is connected to the non-inverting input of a buffer amplifier by an unscreened wire of length 3 m and diameter 0.5 mm. The input resistance of the amplifier is 100 MΩ and its gain is 100.

Two meters away from the connecting wire, and running parallel to it, is another 0.5 mm conductor carrying 115 V at 400 Hz. Assuming that the signal source and the amplifier have a common earth connection, to which the signal wire has a capacitance of 20 pF, and that there are no other significant sources of noise in the system other than that due to capacitive pick-up, what is the r.m.s. value of the noise on the amplifier output? What is the signal-to-noise ratio of the amplifier output in dB?

17.2 **(a)** In an effort to improve the signal-to-noise ratio of the system of Question 17.1, the signal source is connected to the amplifier using a twin-feeder having a capacitance between wires of 3 pF per metre, as shown in Figure 17.1. The twin-feeder is positioned so that its wires are equidistant from, and parallel to, the interfering wire, and 2 m from it. What will be the new signal-to-noise ratio of the amplifier output?

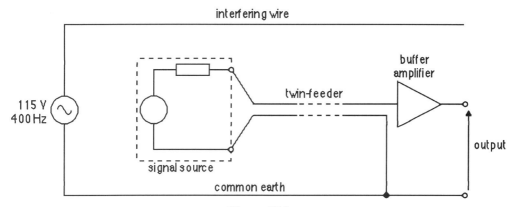

Figure 17.1

(b) As an alternative to the modification described in part (a), the system of Question 17.1 is modified to use an inverting amplifier with the same gain as the buffer amplifier, but having an input resistance of 10 kΩ. What will be the new S/N ratio of the amplifier output?

(c) A third alternative for the system of Question 17.1 is to use the twin-feeder of part (a), positioned as before, to connect the signal source to a differential amplifier as shown in Figure 17.2. The amplifier has a differential gain of 100, a common-mode rejection ratio (CMRR) of 80 dB and an input resistance from each input to earth of 10 MΩ. What will be the S/N ratio of the amplifer output?

Interconnections

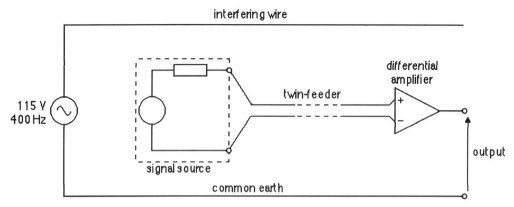

Figure 17.2

17.3 A co-axial cable has a propagation velocity of 240 Mm s^{-1}.

    (a) What is the relative permittivity of the insulation material between the inner core and the screen?

    (b) What is the capacitance per metre of the cable if the inductance per metre is 250 nH?

    (c) What is the characteristic impedance of the cable?

17.4 A transmission line has characteristic impedance 50 Ω, resistance per metre of 50 mΩ and conductance per metre of 0.1 mS.

    (a) What is the attenuation constant of the line?

    (b) What is the signal loss in a 100 m length of cable (i) in nepers and (ii) in dB?

17.5 A signal source is connected to a number of receiving devices by transmission lines as shown in Figure 17.3. The transmission lines have the following characteristics.

line	Z_o / Ω	propn. vel / Mm s^{-1}	length / m
1	50	200	0.5
2	150	240	1.0
3	150	240	0.5
4	300	260	1.0
5	300	260	2.0

The signal source has an internal resistance of 100 Ω and the input resistances of the receiving devices are A: 150 Ω, B: 300 Ω, C: 300 Ω.

    (a) What is the voltage reflection coefficient of each transmission line?

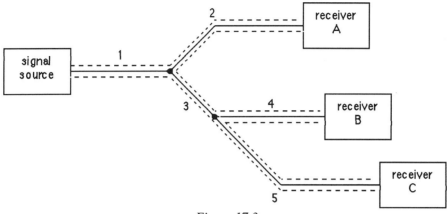

Figure 17.3

(b) What is the VSWR of each line?

(c) If an input pulse of 3.5 V amplitude is generated by the signal source, what is the amplitude of the reflected pulse arriving back at the source?

(d) How long does it take the pulse to reach each of the receiving devices?

(e) What would be a reasonable maximum fundamental frequency for a symmetrical square-wave (i.e. having equal 'high' and 'low' times) if reflected pulses diminish to negligible amplitude after 5 reflections?

Interconnections

SOLUTIONS

Question 17.1

The capacitance per metre between the signal wire and the interfering wire is given by:

$$C \approx \frac{\varepsilon_0 \pi}{\ln(2D/d)} = \frac{8.854 \text{ pF m}^{-1} \times \pi}{\ln[(2 \text{ m})/0.5 \text{ mm}]} = 3.35 \text{ pF m}^{-1}.$$

The total capacitance between the two wires is therefore $= 3 \text{ m} \times 3.35 \text{ pF m}^{-1} \approx 10 \text{ pF}$.

Figure 17.4 shows a noise equivalent circuit for the arrangement.

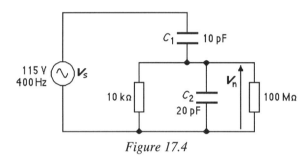

Figure 17.4

The impedance to earth of the signal wire is $\mathbf{Z} = R/(1 + j\omega C_2 R)$, where R is the parallel combination of the 10 kΩ source resistance and the 100 MΩ amplifier input resistance, and is therefore equal to 10 kΩ.

Now
$$\frac{V_n}{V_s} = \frac{\mathbf{Z}}{X_{C_1} + \mathbf{Z}} = \frac{R/(1+j\omega C_2 R)}{1/j\omega C_1 + R/(1+j\omega C_2 R)} = \frac{j\omega C_1 R}{1 + j\omega (C_1 + C_2) R}.$$

For the given circuit values,

$$j\omega(C_1 + C_2)R = j2\pi \times 400 \times 30 \text{ pF} \times 10 \text{ k}\Omega = 7.54 \times 10^{-4}.$$

This is so much less than 1 that we can use the simpler expression

$$\frac{V_n}{V_s} = j\omega C_1 R = j2\pi \times 400 \times 10 \text{ pF} \times 10 \text{ k}\Omega = 2.5 \times 10^{-4}.$$

So, $\qquad |V_n| = 2.5 \times 10^{-4} \times |V_s| = 2.5 \times 10^{-4} \times 115 \text{ V} = 29 \text{ mV}.$

Hence, the r.m.s. noise output of the amplifier $= 29 \text{ mV} \times 100 = 2.9 \text{ V}$.

The peak value of the noise output of the amplifier $= 2.9 \text{ V} \times \sqrt{2} = 4.1 \text{ V}$,

while the peak signal output of the amplifier is $30 \text{ mV} \times 100 = 3 \text{ V}$.

The signal to noise ratio in dB is therefore $20 \log_{10}(3/4.1) \approx -2.7 \text{ dB}$

Solutions

Question 17.2

(a)

The capacitance between the interfering wire and each of the wires in the twin-feeder will be 10 pF, while the capacitance between the two wires of the twin-feeder will be 3 × 3 pF = 9 pF. The equivalent circuit for noise will therefore be as shown in Figure 17.5.

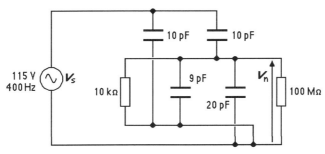

Figure 17.5

Because the return wire of the twin-feeder is connected to earth, its impedance to earth is practically zero, and so the capacitively coupled noise in it will be negligibly small. Also, the additional 9 pF capacitance between the the signal wire and earth is not significant, since we ignore the term $j\omega(C_1 + C_2)R$ in the denominator of the expression for the noise voltage.

The noise voltage at the input to the amplifier will therefore be the same as that obtained in Question 17.1, and the signal-to-noise (S/N) ratio at the amplifier output will remain –2.7 dB.

(b)

With the inverting amplifier instead of the non-inverting buffer amplifier, the noise equivalent circuit becomes that of Figure 17.6.

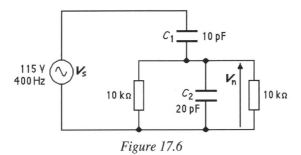

Figure 17.6

The noise voltage at the amplifier input will be

$$|V_n| = |V_s \times j\omega C_1 R| = 115 \text{ V} \times 2\pi \times 400 \times 10 \text{ pF} \times 5 \text{ k}\Omega = 14.5 \text{ mV}.$$

This is only half the original noise input to the amplifier.

However, the signal at the amplifier input is attenuated by a factor of 2 because of the equal source resistance and amplifier input resistance, so the S/N ratio remains –2.7 dB.

Interconnections

(c)

The noise equivalent circuit is shown in Figure 17.7. (I have ignored the capacitances to earth of the wires in the twin-feeder because their effect in the noise calculation is negligible.)

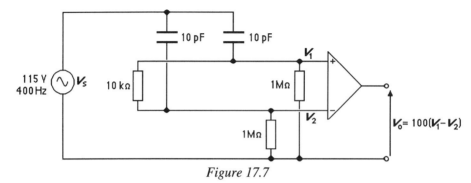

Figure 17.7

The interfering wire will now induce the same noise voltage in each wire of the twin-feeder, so

$$V_{1(n)} = V_{2(n)} = V_s \times j\omega C_1 R = 115 \text{ V} \times j800\pi \times 10 \text{ pF} \times 1 \text{ M}\Omega = j2.89 \text{ V}$$

This noise voltage is present at both inputs to the amplifier and is therefore affected only by the common-mode gain.

Since the amplifier gain is 100 and the common-mode rejection ratio is 80 dB ($= 10^4$), the common-mode gain of the amplifier must be $100 \div 10^4 = 0.01$.

The output r.m.s. noise voltage will therefore be $2.89 \text{ V} \times 0.01 = 28.9 \text{ mV}$, and the peak output noise voltage will be 41 mV.

The new S/N ratio is therefore $20\log_{10}(3 \text{ V}/41 \text{ mV}) = 37.3 \text{ dB}$.

The use of the twin-feeder plus differential amplifier has therefore improved the S/N ratio by 40 dB.

Question 17.3

(a)

Using the equation $\quad v = \dfrac{c}{\sqrt{\varepsilon_r}} \quad$ (equation 17.6 of the textbook, page 657),

$$\varepsilon_r = (c/v)^2 = \left((3 \times 10^8)/(2.4 \times 10^8)\right)^2 = 1.56$$

(b)

Using the equation $\quad v = \dfrac{1}{\sqrt{CL}} \quad$ (equation 17.5 of the textbook),

$$C = \frac{1}{v^2 L} = \frac{1}{(2.4 \times 10^8 \text{ m s}^{-1})^2 \times 250 \times 10^{-9} \text{ H m}^{-1}} = 69.4 \text{ pF m}^{-1}.$$

(c)

$$Z_o = \sqrt{\frac{L}{C}} = \sqrt{\frac{250 \times 10^{-9} \text{ H m}^{-1}}{69.4 \times 10^{-12} \text{ F m}^{-1}}} = 60 \text{ }\Omega.$$

Question 17.4

(a)

For the figures given, R is much less than Z_0 and G is much less than $1/Z_0$, so we can use the equation

$$\alpha = \frac{1}{2}\left(\frac{R}{Z_o} + Z_o G\right) \text{ (equation 17.10 of the textbook, page 658).}$$

and so

$$\alpha = \frac{1}{2}\left(\frac{50 \text{ m}\Omega}{50 \text{ }\Omega} + 50 \text{ }\Omega \times 0.1 \text{ mS}\right) = 3 \times 10^{-3} \text{ neper m}^{-1}$$

(b)

The signal loss in 100 metres of line $= \alpha x = 3 \times 10^{-3}$ neper m^{-1} $\times 100$ m $= 0.3$ neper.

The attenuation $= e^{-\alpha}$ per metre and the total attenuation of x m of line $= e^{-\alpha x}$.

Thus the attenuation $= e^{-0.003}$ per metre, so the attenuation of 100 m of line $= e^{-0.3} = 0.74$.

The attenuation in dB $= 20\log_{10} 0.74 = -2.6$ dB, and so the signal loss $= 2.6$ dB.

Question 17.5

(a)

Transmission lines 2, 4 and 5 are all terminated correctly (i.e. $Z_L = Z_0$) and so the voltage reflection coefficient is

$$\Gamma_v = \frac{Z_L - Z_o}{Z_L + Z_o} = 0.$$

Line 3 is connected to two transmission lines in parallel, each having characteristic impedance 300 Ω, and so is loaded with 150 Ω. So line 3 is also correctly terminated and has $\Gamma_v = 0$.

Line 1 is also connected to two transmission lines in parallel, each having characteristic impedance 150 Ω, so it is loaded with 75 Ω. Since its characteristic impedance is 50 Ω, it is incorrectly matched. In this case,

$$\Gamma_v = \frac{Z_L - Z_o}{Z_L + Z_o} = \frac{75 \text{ }\Omega - 50 \text{ }\Omega}{75 \text{ }\Omega + 50 \text{ }\Omega} = \frac{25}{125} = 0.2.$$

Interconnections

(b)

The voltage standing wave ratio VSWR $= \dfrac{|1+\Gamma_v|}{|1-\Gamma_v|} = 1$ for lines 2, 3, 4 and 5.

For line 1, the VSWR = 1.2 / 0.8 = 1.5.

(c)

Since the reflection coefficient for line 1 is 0.2, the reflected pulse will have amplitude 3.5 V × 0.2 = 0.7 V.

(d)

The time taken for the pulse to traverse line 1 = 0.5 m ÷ 200 Mm s^{-1} = 2.50 ns.

The time taken for the pulse to traverse line 2 = 1.0 m ÷ 240 Mm s^{-1} = 4.17 ns.

The time taken for the pulse to traverse line 3 = 0.5 m ÷ 240 Mm s^{-1} = 2.08 ns.

The time taken for the pulse to traverse line 4 = 1.0 m ÷ 260 Mm s^{-1} = 3.85 ns.

The time taken for the pulse to traverse line 5 = 2.0 m ÷ 260 Mm s^{-1} = 7.69 ns.

The time taken for the pulse to reach device A = (2.5 + 4.17) ns = 6.67 ns.

The time taken for the pulse to reach device B = (2.5 + 2.08 + 3.85) ns = 8.43 ns.

The time taken for the pulse to reach device C = (2.5 + 2.08 + 7.69) ns = 12.27 ns.

(e)

Because the signal source is not matched to line 1, which is itself incorrectly matched to its load, pulse edges will be reflected back and forth along the transmission line with a time of 2.5 ns between each reflection. If 5 reflections are required to reduce the amplitude to negligible proportions, a time of 12.5 ns must be allowed for the received pulse to settle to its new level. A minimum width for each level would therefore be 25 ns (although less time could be used, but with increased chance of error), and the minimum time for one complete cycle is then 50 ns. This corresponds to a maximum fundamental fequency of 20 MHz.